Best Practices in Global Wine Tourism

Edited by

Liz Thach
and
Steve Charters

MIRANDA
PRESS

Best Practices in Global Wine Tourism

Text Copyright © 2016
By Dr. Liz Thach, MW and Dr. Steve Charters, MW, Editors

Published by Miranda Press
An Imprint of Cognizant Communication Corporation
18 Peekskill Hollow Rd P.O. Box 37, Putnam Valley, NY 10579-0037 USA
www.cognizantcommunication.com

Cover design: Lynn Carano Graphics, Inc.

Library of Congress Cataloging-in-Publication Data

Names: Thach, Liz, 1961- editor. | Charters, Stephen, 1957 - editor.
Title: Best practices in global wine tourism / Edited by Liz Thach and Steve Charters.
Description: Putnam Valley, NY : Miranda Press, 2016. | Includes bibliographical references and index.
Identifiers: LCCN2015043576| **ISBN 9780971587069**
Subjects: LCSH: Wine tourism.
Classification: LCC TP548.5.T68 B47 2016 | DDC 663/.2--dc23
LC record available at http://lccn.loc.gov/2015043576

Printed in the United States of America

*Wine cheers the sad
revives the old
inspires the young,
and makes weariness forget his toil.*

-Lord Byron

PRAISE FOR BEST PRACTICES IN GLOBAL WINE TOURISM

"The global breath of this book is impressive, with great case studies from the Old and New World wine regions, as well as emerging wine regions, such as China. I especially enjoyed how each chapter started with a personal story, and the innovative wine marketing tips to promote wine tourism."
- Dr. Janeen Olsen, Author of *Wine Marketing & Sales*

"A very useful book for professionals responsible for implementing wine tourism strategy. The case studies are filled with helpful marketing tips to lure visitors to a wine region or winery, but also describe practical management methods to keep them returning again, and to measure progress."
- Dr. Kyuho Lee, Editor of *Strategic Winery Tourism and Management*

"This book, edited by two of the foremost wine tourism scholars in the world, provides a near perfect mix of academic rigour for students and researchers and practical advice to tourism operators. Each chapter is a case study illustrating practical solutions to real world problems, but the background information, the overview of the regions and countries' industries, and the links to academic literature make this much more than a case book. It is a useful reference for students, a guide for practitioners at the local or regional level, and an intriguing read for wine enthusiasts. The book is separated into two sections; one focuses on regional issues, such as improving the image of an almost forgotten wine region; the other looks at individual winery issues. Each chapter begins with an engaging story or example, which is followed by an overview of the country and region. The case then is presented with a problem, the solution and the best practice details. Highly recommended."
- Professor Larry Lockshin, Professor of Wine Marketing, University of South Australia and Author of *This Little Pinot Went to Market*

"This book stands out from other wine tourism books by its practical focus. A selection of fifteen international best practice case studies provides readers with ample suggestions for successful wine tourism management. The case studies take readers on a journey around the world from old-world appellations Bordeaux and Burgundy, over new-world regions in New Zealand and Argentina up to emerging markets such as China. The cases will inspire both wine education and wine practitioners. I recommend this book to anyone who wants to learn how wine regions and wineries successfully solved actual problems in wine tourism."
- Dr. Simone Mueller Loose, Professor of Wine Marketing, Geisenheim University,

"An excellent collection of fascinating and useful case studies describing the challenges and triumphs of implementing a wine tourism strategy. An engaging read for wine hospitality managers, enthusiasts, and students."
- Dr. Don Getz, Author of *Explore Wine Tourism*

Contents

Chapter One

Introduction and Overview of Wine Tourism

Steve Charters and Liz Thach

ESC Burgundy School of Business, France & Sonoma State University, California

For centuries tourists have enjoyed sampling different wines of the world during their travels, but the academic study of wine tourism is only of a comparatively recent vintage. The first studies were carried out in Texas, by Tim Dodd in the mid-1990's, yet in the last twenty years research in the area has grown dramatically and it is now a recognized aspect of tourism research, as well as a significant element of wine marketing. The fact that it covers both of these fields, as well as hospitality, geography, and management, is part of its appeal, as it brings together different ways of working, varying methodologies, and often divergent perspectives.

This volume is designed to reflect that breadth of disciplinary background and multiple ways of viewing the topic, ranging from attracting tourists to wine regions in the backroads of Greece, Argentina, Tasmania, and China, to finding new methods of wine tourism management in Provence and Sonoma. The time is also right not just for further research, but to expand the repertoire of resources to those who teach on the burgeoning number of wine tourism courses in the world.

Therefore, the purpose of the book is not merely to improve academic understanding of wine tourism: the hope is that the cases featured here will also be useful to practitioners in the industry. An examination of best practices and examples will, we hope, be a stimulus to those working in the field that are seeking to manage their wine and/or tourism businesses more efficiently, manage their regional bodies more effectively, and understand the needs and expectations of consumers more clearly.

DEFINITIONS AND TYPES OF WINE TOURISM

There have been a number of definitions of wine tourism. The first comprehensive one, and perhaps still the most widely accepted, is from Michael Hall (1996, 2000). This suggests that wine tourism comprises:

"Visiting vineyards, wineries, festivals, and wine demonstrations to taste grapes/wine and/or to experience the attributes of a wine region being the main reasons for the visit."

Donald Getz (2000, p. 4) expands this definition to include a broader audience:

"Wine tourism is travel related to the appeal of wineries and wine country, a form of niche marketing and destination development, and an opportunity for direct sales and marketing on the part of the wine industry."

Both definitions are very helpful, however Carlsen (2004) suggests that a single agreed definition is yet to be established. Notably, Hall's definition assumes that wine is the main purpose of a tourism visit; however, the goals of a tourist may not be easily separated in that way and it could be better to suggest that wine is one of the reasons for a visit. Additionally, there are both supply and demand sides to be considered, and the supply side itself comprises wine producers and tourism operators of all kinds. Additionally, regional authorities, responsible for tourism and/or economic development, may also have a valid interest in the activity – and in Europe, at least, may be the dominant actors in developing it.

We can note that wine tourism comprises a number of activities, so that the tourists can explore a wine region, they can taste and buy wine, seek education about how wine is made and where grapes are grown, and explore the relationship of wine and vineyards to the local environment, and the history and culture of the area (Carlsen, 2004). In this way, wine tourism comprises two types of activity. There are those which are immediately linked to the consumption of wine, and those which are indirectly linked to wine. The former includes not just cellar door visits, but also events and festivals, tours and educational activities, such as guided visits in a winery or vineyard. The latter links wine to other attractions and to general hospitality in a region. Landscape walks and heritage attractions, art and craft centers, restaurants, and bars are all in this category, and are as essential to the visitor as the cellar door or the vintage festival.

THE BENEFITS AND CHALLENGES OF WINE TOURISM

Wine tourism in New World countries has become important. California, for example, received over 20 million visitors at its wineries in 2012 and they spent over $2 billion (Thach, 2013); this has inevitably had a noticeable impact on the state's economy. Wine tourism is also a key income generator in New Zealand and Australia; in the former country 13% of all international visitors visit at least one winery, making 220,000 visits per year. They are high-value tourists, spending 32% more on average during their stay than the average tourist (New Zealand Tourism, 2014). In Australia, it has been estimated that there are more than five million visitors to the country each year that go to a winery during their stay (SA Tourism Industry Council, 2009).

Beyond the pure financial benefits to a country, however, three other broad sets of benefits can be seen. The first is to a region which includes an area producing wine. Beyond the mere value of tourist spend, wine can be a major part of the destination image of a region. Thus, in Champagne, wine blends with French culture, gastronomy, heritage – such as Reims cathedral – and a dynamic history with a particular focus on the First World War (Menival & Charters, 2011). In terms of the overall attraction, the region wine cannot easily be disentangled from the other factors, yet it remains fundamental to how the image of the region is constructed in the minds of those who visit it. This, in turn, translates into a wide array of potential economic benefits – to the hospitality industry, tour and travel operators, and in taxes accruing to public authorities. Nevertheless, the management of this territorial brand is not necessarily easy, requiring as it does substantial clarity of vision and cooperation between actors (Charters & Spielman, 2014) and, especially in Europe, can often become bogged down in administrative conflict and confusion (Correia et al., 2004).

The second and third benefits accrue directly to wine producers. Most obviously, there is the chance to enhance sales, and especially gain sales direct to the consumer, thus cutting out distributors and retailers, which optimizes the profit to the producer. This is the main reason why winery owners decide to become involved in tourism. At times it may be the failure to sell their wine quickly enough which stimulates activity at the cellar door. The negative aspect of this is that because it is sales driven, tourism activities are not always carefully planned, and the offer may lack investment compared with what is put into the production side of the business. The more important, but often overlooked, benefit is the development of the winery's image, and the growth of brand equity. This is harder to measure, but is likely to have greater long term benefit, with the possibility of continued loyalty, repeat sales over many years, and a number of word-of-mouth recommendations to friends who themselves may become visitors and/or regular customers.

Major Benefits & Challenges of Wine Tourism

Benefits	Challenges
✓ Enhanced Revenues for Region ✓ Enhanced Regional Image ✓ Increased Cellar Door Sales ✓ Improved Winery Image ✓ Brand Growth & Recognition	✓ Collaboration ✓ Regulatory Approvals ✓ Environmental Issues ✓ Traffic Congestion ✓ Financial Support ✓ Measurement

The major challenges of wine tourism have to do with the need for strong cooperation and partnership between the various entities involved in making it a success. Without restaurants, hotels, police, medical agencies, local government for permits, environmental groups, employment services, road agencies, and other infrastructure support, wine tourism cannot succeed (Thach, 2007). Neighbors must be consulted about increased traffic and potential noise pollution. Financial support must be obtained, and a marketing campaign designed which includes not only brochures and a website, but signage, event planning, and evaluation. Most successful wine tourism efforts invest in a regional branding program and elect a board of directors to guide the process. Additionally, care must be taken to protect the environment and rural beauty of the vineyards so that the region maintains its charm and reason to be a wine destination. Finally, it is often challenging to establish a measurement system to verify that efforts are useful and there is a reasonable return on investment.

TYPES OF WINE TOURISM RESEARCH

Research into wine tourism so far has mirrored this split of activities. There can be a focus on the territorial brand of a wine region (Charters & Spielman, 2014), on the tourism offering and on wineries. Each of these have a supply side aspect, considering how the offer is managed and its value, and each a demand side perspective, considering how consumers respond. This is especially the case in our era of the experience economy (Gilmore & Pine, 2007) where it is important to examine the tourists' overall experience of the visit, or event, or tour. Within this form of experiential tourism, aspects of expectation and satisfaction have been significant, and examined in some detail.

WHAT RESEARCH TELLS US ABOUT WINE TOURISTS

At the outset, it is important to observe that most academic research into wine tourism has been carried out in New World countries, where the phenomenon has been more clearly delineated over the last 20 years. The result is that our understanding of the subject is skewed by this geographic research focus, and while in Italy especially, there have been more European studies over the last decade, it does mean that how we view tourists comes with an Anglo-Saxon bias.

It is also significant that research suggests that the characteristics of wine tourists vary from country to country and, indeed, from region to region. Connoisseurs, who in Europe may join a carefully organized tour led by a Master of Wine, are more likely to construct their own tour in South Africa or New Zealand, contacting wineries directly. International visitors to Australia are most likely to visit the Hunter or Barossa Valleys, because of their proximity to metropolitan centers, rather than the equally renowned Margaret River or Coonawarra regions, more distant from population centers.

Additionally, most visitors to wine regions are more highly-involved drinkers. Many people enjoy wine, but would not think of visiting a winery when on holiday, and even less of making a special trip to one. These are also the consumers who are less interested in learning about the detail of the product or in exposing their uncertain tasting ability in a public environment. The result is that wine tourists have typically been segmented into categories based on interest, knowledge, and regularity of consumption. Labels such as wine novices, wine interested, or wine lovers are regularly adopted for the purpose of segmentation.

Crucially, what study after study has shown is that the wine tourist is part of a wider category of experiential tourist, often directly linked to food tourism, as well as sometimes to agri-tourism, ecotourism, and heritage tourism. This experiential aspect is important because it can be contrasted with the idea that the tourists are overwhelmingly focused on the wine. Indeed, while tasting and buying wine remain high on the agenda, the need to have a good time or socialize with friends has been observed regularly from over 15 years ago (O'Neill & Charters, 2000) to the present day (Bruwer et al, 2014). This is very important, as it remains an *idée fixe* of producers that as long as their wine is good, they will be able to sell it. This fails to understand that what most of their visitors want is not the very best wine, though they will reject poor quality wine, but an enjoyable time, in a pleasant environment, with staff who make them feel welcome (Thach & Olsen, 2006).

OVERVIEW OF MAIN WINE TOURISM REGIONS, AND EMERGING WINE TOURISM REGIONS

It is probable that if the highly-involved wine tourist was asked to name the world's key destinations then they would start by listing European, and particularly French, areas. Bordeaux, Burgundy, and Champagne would be high on the list, perhaps together with the Northern Rhone, Chateauneuf-du-Pape, and Alsace. To these may be added the Rheingau and the Mosel Valley, probably Rioja and Jerez, the Douro Valley and in Italy Tuscany, particularly, with Piemonte a close second.

Nevertheless, as has been implied before, wine tourism in the New World is often more organized than in Europe. In the Old World, some of the most reputable regions feel that their wines sell well enough without needing to develop their image by welcoming visitors, even though the challenge of innovative wines from New World wine countries from the 1990's onwards has shaken this. Also important is the fairly widespread viewpoint, at least some parts of Europe, that when they receive visitors, wineries are merely selling wine, and not engaging in wine tourism (Menival & Charters, 2011). This limits the offer, and can reduce the experiential value for many visitors.

Major Wine Tourism Countries

Old World	New World	Emerging	Budding
France	California	Croatia	China
Spain	Australia	Greece	Thailand
Italy	New Zealand	Austria	India
Germany	South Africa	Hungary	Japan
Portugal	Argentina	Canada	Korea
	Chile	Romania	Vietnam

It is therefore in New World regions that one often sees the most well-developed and innovative forms of wine tourism. Important here are the Napa Valley, and other Californian wine areas such as Sonoma and Santa Barbara, Marlborough in New Zealand, Stellenbosch and Constantia in South Africa, Mendoza in Argentina, many of the Chilean regions close to Santiago, and those around Adelaide in Australia, as well as the Yarra Valley near Melbourne, and especially, the Hunter Valley near Sydney.

However, as the world of wine is becoming ever more international, so the world of wine tourism is producing some other, less well-known, but increasingly popular areas. Santorini in Greece, with its assyrtiko grape, the Wachau in Austria, especially with gruner veltliner, the Niagara region in Canada, and the Tokaj region in Hungary all fit into this category. To these can be added many lesser-known regions in the traditional European leaders, where a lower reputation for their wines and increased pressure on sales is forcing more attentiveness to and innovation in wine tourism. Examples here could include the Alto Adige or Sicily in Italy, much of the Languedoc in France and Galicia or Priorat in Spain. More recently wine tourism has begun to bud in the Far East, with wineries in China, Thailand, India, Japan, Korea, and Vietnam starting to promote wine tourism offerings (Asian Wine Association, 2015).

FUTURE ISSUES TO EXPLORE

The foregoing has already highlighted areas within wine tourism which require more attention. Clearly further work is needed in Europe, to assess the effectiveness of what is provided in both famous and less well-known areas, and to assess the expectations of consumers and see how they may differ from the current, predominantly non-European, knowledge. Crucially, in Europe there is a major focus on the concept of terroir, and the use of appellation systems to underpin it, and the relationship of these ideas to tourism. In particular, the notion of authenticity is not well understood.

At the same time, cross-cultural studies, examining the expectations, experience, and satisfaction of visitors from different cultures, could be very helpful for the wine industry in the future. Most obviously, this could compare French or Italian visitors in their native land with Anglophones in the same place. An even more important research question is what does the new wave of wine drinkers, especially those from various East Asian countries, seek from a wine tourism experience? This is especially important given that they often have different ways of structuring their travel.

Not enough attention has been given to the territorial brand in wine tourism. How does it relate to the idea of a cluster? How can it be better managed? How can it develop a shared vision and mythology? In more *dirigiste* cultures how can a wine tourism plan win a sufficient constituency amongst the individual businesses to avoid being merely another form of public administration? All of these are significant questions which can help to improve the wine tourism offering.

Destination image also has had little attention, especially in discrete wine regions. What are the components of destination image, and how do they relate to the wine itself? In this, regard the relationship of wine reputation to destination reputation deserves much more research. If the image of a wine is poor, or declining, how does that impede wine tourism in a region? Can an innovative and enjoyable wine tourism offer help to remedy a wine of low standing?

Finally, almost all research on wine tourism has been carried out in fairly well-established areas. An interesting focus for further study would be to take a region whose vinous reputation is very new, or emerging from a period in the shadows, and to explore the relationship between the wine and the tourism in the region. Places such as the Jura in France, Puglia in Italy, or Swartland in South Africa, may fit into this category.

Overview of the Book – Best Practice Case Studies in Wine Tourism

Given the issues identified about the motivation and effect of wine tourism, it should be evident that there is much still to learn about what wine tourists are searching for and how wine tourism can be most effectively delivered. This book, therefore, aims to fill at least some of those gaps in our knowledge, and provide practitioners with some ideas on how they can improve the organization, coordination, service delivery, and creation of an outstanding experience for their visitors.

In focusing on **"best practices"** in wine tourism, we have been guided by the standard definition that "a best practice is a method or technique that has shown results superior to those achieved with other means" (Business Dictionary, 2015, p. 1). At the same time, we recognize that a best practice can evolve over time as improvements are made (Bogan & English, 1994), and therefore it is probable that many of the wineries and regions featured in this book will continue to adapt and improve in the future.

As we searched for best practice case studies to include, we relied first upon the expertise of the authors to identify either regional examples or specific winery examples from their country. Our second concern was to attempt to find at least one case study from the major wine countries or continents of the world. In this way, we were primarily successful in that there is a case study from each of the major continents with the exception of Africa. However, certain countries, such as France, appeared to have a wide variety of examples, and it is for this reason that there are four chapters on France, highlighting best practices in the regions of Burgundy, Bordeaux, Provence, and Beaujolais. In the end, we found fifteen excellent examples of best practices in wine tourism from around the world.

Flow of the Book

In the first half of the book, we focus on regional wine tourism where best practice solutions are implemented to solve a problem within a specific wine region, rather than with an individual winery. For example, in the second chapter on Austrian wine taverns in Vienna, the local tourism agencies along with wine tavern owners worked together to lure tourists back to the traditional wine bars called "heurigens." In the third chapter on Niagara, Canada, wineries grappled with the issue of how to attract tourists in the dead of winter, and created the very successful Winter Wine Festivals. In the fourth chapter, wine industry officials in the Ningxia region of China struggled with a very different issue of trying to attract tourists to a brand new wine region.

Chapters five though seven highlight best practices in the regions of Burgundy, Bordeaux, and Provence, and emphasize how important it is even for well-known wine regions to continually improve and refresh the tourism offering. Chapter eight tells the story of how Italian wineries worked together with the government and regional associations to create the very successful *Citta del Vino*, or cities of wine, program throughout Italy to attract tourists.

Chapter Nine describes how the isolated island of Waiheke in New Zealand managed to attract more tourists during off season and also to create more attractive infrastructures for tourists to visit and move around the island. Chapter Ten describes the trials of the Sonoma wine region in California, that used to be dwarfed by the success of neighboring Napa Valley, and through collaborative partnerships with the various tourist organizations managed to attract more tourists and revenues, as well as to win an outstanding wine tourism region award.

The second half of the book is devoted to best practice stories of individual wineries. This begins with the intriguing case of Zuccardi Winery in Mendoza whose location, off the regular tourism trail, was causing problems until they implemented several innovative programs to attract more tourists. Chapter twelve describes similar struggles with Moorilla Winery in Tasmania, Australia, and explains the many unique wine tourism experience they created to become a top wine destination in the country.

If you've ever wondered why wineries don't focus much on children as part of the tourism offering, Chapter thirteen, which describes the Hameau Duboeuf Winery in Beaujolais, will explain best practices in making tourism work for families. The next stop is mainland Greece where Gerovassiliou Winery provides an excellent example of how to be distinctive by focusing on history, education, and reviving a lost wine heritage. They too, have found a formula to include the whole family, as well as school children, in the wine tourism equation.

The final two chapters are set in Portugal and Spain. The case of Herdade da Malhadinha Nova Winery in Portugal is intriguing in that, again, the family battles to attract tourists to the Alentejo region, which is not very well-known. They succeed by creating an amazing smorgasbord of wine, culinary, and adventure experiences that have made them a top draw in Portuguese wine tourism. Likewise, Bonastre Winery in the Catalonia region of Spain, is beset with issues when the global economic recession turns into a crisis in Spain. Fighting to stay afloat, the owners develop some innovative event offering that not only allow them to keep their business, but to become one of the top tourism stars in the region.

In addition to learning about the specific wine tourism problems and solutions used by each winery or region, the reader will also learn more about the wine industry in each country, including basis statistics. At the end of each case, there is an explanation of results and why it is considered to be a best practice. In addition, future issues are identified and discussion questions are provided for those who chose to use this book for educational purposes.

We hope you enjoy this virtual tour to some of the most famous wine regions of the world, and a few of which you may never have heard. This, of course, begs the question about whether "virtual wine tourism" should be included in the definition of wine tourism. Something to reflect upon as you read this book, and sip an enticing glass of wine. Cheers!

A Toast from Wine Tourists

11

Regional Wine Tourism
Best Practices

"Wine is one of the most civilized things in the world and one of the most natural things of the world that has been brought to the greatest perfection, and it offers a greater range for enjoyment and appreciation than, possibly, any other purely sensory thing."

— Ernest Hemingway

Chapter Two

Luring Tourists Back to the Traditional Wine Taverns (Heurigens) of Vienna, Austria

Albert Stöckl & Cornelia Caseau

University of Applied Sciences Krems, Austria & ESC Dijon, France

It was a warm September afternoon when Franz, a regular customer, entered a *heurigen*, or wine tavern, in Vienna, Austria. Franz was not only a neighbor and patron of the heurigen, but also one of Austria's more than twenty thousand regional grape growers (Austrian Wine, 2015). Franz sold his small amount of grapes to the heurigen owner every year, but he was not happy. Kilo prices for grapes had gone down again, and soon it would not be profitable to toil all year in the steep rows of vines of his one hectare vineyard located on Vienna's famous Nussberg.

When Franz complained, the heurigen owner argued that wine sales had fallen and his heurigen customers were drinking less each year. Because of this, he couldn't afford to pay higher kilo prices, even though the quality of Franz's grapes was excellent.

"Customers do not worship my traditional wines and food any more", the elderly tavern owner whined. Franz sighed heavily, and after finishing the last drop of wine in his glass, placed it carefully on the wooden table and walked to the door.

"Wer nicht mit der Zeit geht, geht mit der Zeit," he grumbled as he left.
(Translation: "He who does not move with the times will be removed over time.")

Franz was not the only one dissatisfied with the low price of Austrian wine grapes. Even though Austria had recovered from the 1985 scandal when some winemakers had been caught adding diethylene glycol to their wine in order to sweeten it, which caused most countries to pull Austrian wine off retail shelves (Tagliabue, 1985), the wine situation in Austria was still not bright. Many local consumers were beginning to seek out other beverages besides wine, and international consumers were confused over difficult sounding Austrian grape varietals such as gruner vetliner and zweigelt.

Even more distressing was the plight of the Austrian wine taverns – the heurigens – that traditionally had served in roles of both restaurant and retail store for local wineries. Many people accused the heurigens of not keeping up with changing consumer needs, such as desires for low versus high calorie diets, reduced consumption of alcohol, and healthier organic food and beverages.

While the heurigens strived to maintain tradition, authenticity and local customs, customers – especially the younger ones – were clamoring for change. This case study describes how a positive partnership between wineries, heurigens, the Austrian Wine Marketing Board, and the Vienna Tourist Board worked together to revamp the image of heurigens and lure tourists back to enjoy Austrian wine. It is based on in-depth interviews with ten experts who worked in these various establishments.

OVERVIEW OF THE AUSTRIAN WINE INDUSTRY

The history of wine in Austria goes back to 276 AD when the Roman emperor Marcus Aurelius (in Austria also called "Probus") cancelled an edict that forbade wine growing outside of Italy, in the Roman provinces. Soon after, vineyards were planted in many regions of Austria, and today Austria has 45,000 hectares of winegrapes (Statistik Austria, 2013), producing an estimated 2.25 million hectolitres of wine in 2014 (USDA, 2014). There are approximately 6,500 wineries in Austria (ÖWM, 2010).

# of Wineries in Austria	6500
# of Wineries in Vienna	630

More than 65% of the wine produced is white, with grüner veltliner as the signature white grape at 29% of production (USDA, 2014). This is followed by welschriesling and mueller thurgau. The most popular red varieties are zweigelt and blaufraenkisch. The Austrian government is trying to promote organic farming, with 10.5% of vineyards using organic management methods in 2012 (USDA, 2014).

Wine production is focused in the east and south-eastern sections of Austria. There are four major wine producing regions in Austria (Austrian Wine, 2015):

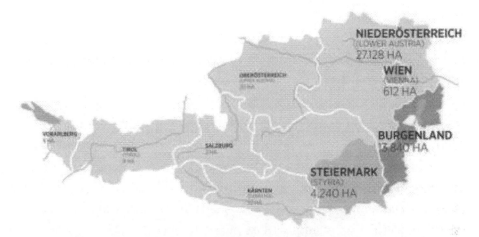

Map of Major Austrian Wine Regions

1) *Niederösterreich* – This is the largest wine region at more than 27,000 hectares. It is known for producing high quality wine with a focus on grüner veltliner. It includes famous and very popular subregions such as the Wachau and Kamptal. There are many tourist trips along the Danube River to visit the wineries in this region.

2) *Wien* – Vienna is the only capital city in the world that has active vineyards within it. Today, vineyards cover around 1700 acres of Viennese' hillsides and there are 630 producers (Gross, 2014). The outskirts of Vienna all have vineyards, but 87% are located in the hills of the 19th and 21st Viennese districts in the north-west where they form the right and left bank of the Danube. This is also where the famous wine and heurigen villages like Grinzing, Neustift and Stammersdorf are situated.

3) *Burgenland* – Located south of Vienna, this region is warmer and specializes in red grapes, such as zweigelt. It is also home to the famous Lake Neusiedl, and the charming tourist town of Rust, where storks nest on rooftops and taverns serve Burgenland's liquid gold – a unique dessert wine called Ruster Ausbruch.

4) *Styria* – This southern most wine region of Austria, located to the north of the country of Slovenia, is internationally known for its unique sauvignon blanc. Due to its more mountainous typography and higher altitude, the climate is cooler and they produce light and elegant wines of varietals such as muskateller, sämling and traminer. A major tourist attraction here is visiting the famous wine-route of southern Styria.

Austrian wine exports have grown over the years, with 50 million liters of wine, worth 146 million euros, exported in 2014 (Austria Wine, 2015). Top export countries are Germany and Switzerland, but wine exports continue to grow in other markets such as the USA, China, the UK, and Scandinavia. The average price of Austrian wine has also increased, based on the Austrian wine industry strategy to move from jug wine to higher quality production.

Wine Tourism and the Importance of the Austrian Heurigen

Austria is the 8th most visited country in Europe, with more than 24.2 million international tourists in 2014 (UNWTO, 2015). According to Vienna.info (2009), tasting Austrian wine and visiting heurigens is the second major motive for tourists who come to Austria, with the first being to visit coffeehouses.

The heurigen, Austria's and, in particular, Vienna's traditional wine taverns, have played and still play an important role in both the local wine and food culture and in wine tourism (Keen & Robinson, 2001). A heuriger can be defined as a tavern that offers wines from the current vintage, as well as simple food in a cozy, warm, and welcoming environment (Wein.Info, 2014). The term, "heuriger," literally means "this year's", and has three distinct meanings: young (this year's) potatoes, this year's wine, and the traditional wine tavern where the latter is sold (Robinson, 2006).

The history of the heurigen is interesting, and began in 1784 when Austrian Emperor Joseph II, son of Maria Theresia and brother of Marie Antoinette, created a law that allowed winemakers to open their facility, which was usually the stall or wine cellar in those times, to the public to sell their own wines. The particularity of Joseph's law lies in a small but significant detail: the sale of wine made from one's own produce was and still is quasi-free from taxes. Small wineries in Austria are taxed on a flat rate basis and wine and food sales at the heurigen are included in these "all-in" regulations.

After World War II and through the 1970's, the estimated number of heurigens in Austria was at an all-time high of around 20,000. At that time, most heurigens operated in a very simple format, often in small stalls or converted living rooms. Heurigen owners served their young wine along sparkling water, and many guests brought their own food. In those days, the heurigens were by far the most important wine distribution channel in Austria. Today, the current estimate shows approximately 2,000 heurigens in Austria (Pleil, J., personal communication, August 2015).

Food and wine in a traditional heurigen are quite simple. The young, dry, and mostly white wine is accompanied by bread and hearty food such as Liptauer, a cheese spread made of soft cheese, paprika, and spices. In addition, cold, thinly cut pork roasts such as Schweinsbraten, Kümmelbraten and Surbraten, as well as sausage products such as black pudding or liver sausage, are served.

Guests Enjoying Wine at a Heurigen

In addition to selling local wine and food, the heurigen has an important cultural function. In the tavern, everybody sits together to eat, drink, and talk, regardless of socioeconomic status (Kopsitsch, 2008) and is treated in the same friendly but informal way by the landlord and his family or staff.

The tables are generally big, simple wood tables without tablecloths. Instead of chairs there are wood benches. Very often strangers share a table, which is for six to ten people. Doctors, lawyers, or managers sit side by side with workers, students, pensioners, or tourists.

Music is also an aspect of the heurigen, with Schrammelmusik played by three to four musicians using violin, button accordion, and contraguitar. The melancholic lyrics of the songs mostly deal with drinking wine, fun in life, and, ironically, death. The Heurigen songs are said to express the "… whole sweet carefree Viennese love of life and at the same time a deep despair to have to be wiped out some time," (Sinhuber, 1996). This type of music can, if at all, only be compared to the Fado music of Portugal, and constitutes an important part of Viennese musical identity and self-conception.

THE PROBLEM – STRICT LEGAL REGULATIONS AND CHANGING CONSUMER EXPECTATIONS

In the past, the heurigen wine taverns served mainly as a profitable way of commercialization for winemakers (Wieninger, 2013). This was facilitated by the fact that the wine producer was also the wine tavern owner, or the person running the heurigen belonged, and still mostly belongs to the same family. For a long time the prevailing rule for Austrian wines was quantity before quality, but after the infamous wine scandal in the eighties, a reorientation became necessary and customers had to be reassured. Eventually the heurigens were able to overcome consumer distrust, mainly by personal connections and reassurance from the heurigen owners.

But by the early 2000's another looming issue threatened the future of the Austrian heurigen, and this was a change in consumer perceptions and desires regarding wine and food quality, and restaurant ambience. This, coupled with onerous legal regulations regarding the type of food and wine they could serve, along with an increase in Austrian wine exports, put huge pressure on heurigens. Many were forced to close their doors because they could no longer make a profitable living. The following issues highlighted the problems.

Strict Legal Regulations to Operate a Heurigen - Partially because of their tax free status, the Austrian government put strict regulations on how heurigens could operate. A heurigen can only sell cold food and locally made wine. They are not allowed to sell hot food, coffee, beer, non-traditional soft drinks such as Coca-Cola or Fanta as well as purchased (not home-made) wines or spirits. A real heurigen is therefore very limited in its actions in order to respond to consumption trends. Neither is it allowed to offer wine that was bought elsewhere, nor does the heurigen manager have the right to provide non-traditional foods such as sushi, Spanish style tapas, or Italian antipasti, even if they are only cold dishes. Furthermore, a heurigen does not have the right to be open all year round. Depending on the region it can only be open from two to four weeks in a row and a limited number of weeks per year.

Desire for Higher Quality Wine – The Austrian wine scandal woke up consumers to the need to consider quality and source of wine, and soon it became one of the major criteria for choosing their wine. Consumers began to look for quality designations on wine bottles, such as wine of certified origin and quality. Imported wine also became more popular in Austria. This change was challenging to heurigens as traditionally they only offered young wine from their estate that was not served in a bottle, but in a simple jug or carafe.

Focus on Healthier Food – At the same time consumers were asking for high quality wine, many started to look for fine gastronomy or low-calorie products, such as salads, instead of fatty roasted pork in the traditional taverns. Also public campaigns about a healthier diet, the vegan, vegetarian and organic trends, the ecological and the Slow Food movement and sustainability claims emphasized this change in eating habits. Since the law restricted heurigens from offering many of these types of cuisines, their situation was made even more challenging.

Stricter Drunk Driving Laws – In the past the traditional heurigen buffets were substantial and helped to neutralize the high consumption level of alcohol. However, stricter regulations and frequent security controls of the police in the streets also shaped the consciousness of the consumers. Many consumers stopped drinking as much wine as they had in the past.

Young People Want Hip Establishments - Young consumers started drinking wine in more stylish establishments, such as modern bars, restaurants, and *vinothèques* in the city center. The heurigens, with their dated décor, older clientele, and family orientation, were not as attractive to the younger crowd.

Winemakers Exporting More – In the past, Austrian wine was essentially a locally sold product, with the majority of it sold in heurigens. Today the top winemakers export a third of their production to countries such as Germany, the USA, Japan, the Netherlands, Finland, and Switzerland, but also, in smaller proportions to reputed wine producing countries like France, Italy, and Spain (Bachmayer, 2012).

THE SOLUTION – RESEARCH, COLLABORATION, & INNOVATION

Fortunately, the fate of the heurigens soon became a public issue and a source of concern for the Austrian wine industry. After all, they were a unique part of Austrian culture and were still important to international tourists, especially in the city of Vienna, where the majority of the heurigens were located. Therefore, the Vienna Tourist Board, the Austrian Wine Marketing Board, winery/heurigen owners, and university researchers began investigating the issue. Though it wasn't a formal partnership, all parties made efforts to solve the problem, and the following steps were implemented.

Step One: Investigative Research

A group of university researchers began investigating the issues associated with heurigens and published their findings. Strobl (2002) studied 150 heurigens in Vienna and developed a classification system based on level of development. She found that some heurigens had obtained to be a full restaurant, and though they lost the tax advantage, they benefited from being able to offer other beverages and food. She also verified the critical importance of the heurigen as a distribution channel for Austrian wine.

Baumgartner (2004) took a more in-depth look at traditional heurigens verses those that had become restaurants and realized that the marketing tactics used by each were quite different. Whereas the traditional heurigens relied on word-of-mouth and attracted an older clientele, the more modern tavern style restaurants utilized the Internet and other modern methods of marketing communication that attracted tourists and a younger clientele. She discovered the modern taverns used trained staff, whereas the heurigens, which were run by the owner and family. Also an advantage to being a modern tavern was the ability to stay open more days of the year and offer more diverse types of cuisine and beverages, which was attractive to tourists.

Schiener (2007) interviewed 279 guests visiting 13 regional heurigens and discovered they were mostly locals from the age of 40 to 60, and came to listen to music and/or smoke. Research conducted by Beiglböck (2007) highlighted the need for heurigens to attract young people and tourists in order to survive in the long run. More recently, a series of semi-structured interviews were conducted with key stakeholders involved in creating some of the strategic changes to help revamp heurigens (Stöckl et al, 2014). The results of these interviews were used to identify the best practices described here.

Step Two: Development of New Marketing Strategies

Eventually, some of the heurigen began to adopt new marketing strategies in order to lure new consumers to their premises. A key aspect of this was clear branding, with the owner's name linked to the reputation and high quality of wine and food offered by the heurigen. With this change, more consumers began to make their choice based on the name of the tavern owner. Strongly branded heurigens such as *Wieninger, Mayer am Pfarrplatz, Edlmoser,* and *Christ* became more popular. In this way, the name of the heurigen owner became a quality guarantee of a strong brand, which could be transferred to new generations (Bachmayer, 2012).

To support the branding efforts, high quality marketing communication was adopted by some of the heurigens. They began to create webpages to show photos of the heurigen and illustrate the family connection to the land and high quality of the wine. They also provided an online menu and began to adopt some social media tools, such as Facebook and Tripadvisor, as well as smart phone apps to provide real-time information regarding opening hours, location and menu updates. Heurigen owners began to send out newsletters to stay in touch with customers, and many started to participate in wine competitions and communicate medals and awards they received for their wine.

Step Three: Link to Environment

Since appreciation for terroir and preserving nature has always been part of the philosophy of heurigens, many began to communicate their relationship to the land and the methods they used to be environmentally conscious. For ecologically minded consumers they communicated the fact that most of the heurigens were easily accessible by public transport. Heurigens that were using organic or sustainable farming and production techniques began to emphasize this more in their marketing, as well as recycling efforts in production and packaging. Since the heurigens are mostly situated in districts with large green spaces or on the foothills of the Viennese woods, some of them started organizing wine hikes through the vineyards and neighbourhoods to attract tourists.

Step Four: Powerful Partnerships

Fortunately the Vienna Tourist Board and the Austrian Wine Marketing Board also recognized the importance of the heurigens to both Austrian culture and wine distribution. Therefore they began promoting heurigens to international tourist groups, and communicating about the concept of heurigens on their websites and international marketing trips. The Austrian Wine Marketing Board invites journalists and other media to visit heurigens when conducting wine tours of the country. In addition, they began to develop campaigns to assist consumers in pronouncing some of the more challenging Austrian grape varietals, such as "groovy" for grüner vetliner (Teague, 2009).

Step Five: Innovative Thinking

During this period, several examples of innovative thinking emerged. Probably one of the most intriguing was the development of a new wine style called *Gemischter Satz*. Based on an ancient Austrian field blend recipe, the wine must be composed of at least three types of grapes from one Viennese vineyard. The main type of grape must not exceed 50%, and the third type of grape must amount to at least 10% (Wienwein.at). This variety only exists in Vienna, and now almost every traditional heurigen offers at least one own version. In the last years, *Gemischter Satz* has become a very trendy product, appreciated by all types of wine-drinkers. At the same time, many producers began to craft wines that were lighter in style, and also adopted more organic and sustainable methods. More winemakers began to investigate biodynamic winemaking principles, created by their fellow Austrian, Rudolf Steiner.

Wien Wein Group

Another example of innovative thinking was the collaborative partnership called the *Wien Wein Group,* where six young wine estate and heurigen owners decided to work together to promote Viennese wine and wine taverns. They created a website called *wienwein.at* as well as brochures that emphasize the uniqueness of their wines and their pride in Austria. They use intensive personal branding in presenting themselves in a flattering way in different places around Vienna, or in doing their job in the vineyard, in the winery or in the wine tavern.

Wien Wein focus on attracting younger, more urban consumers as well as international tourists to their heurigens, and use modern digital marketing methods to do so. In addition, they have made a clear effort to communicate the double orientation and connection of their role as wine growers and heurigen owners. For example, the logo of the Wieninger brothers (Fritz Wieninger is the winemaker and his brother Leo the heurigen manager) is composed of the skyline of Vienna, discovered by chance by Fritz on a postcard some years ago, followed by their family name. The logo addresses not only the local wine consumer, in using the famous skyline with Vienna's most well-known emblems, St. Stephen's cathedral and the Giant Ferris Wheel, and can easily be remembered by wine drinkers who come from outside Vienna.

Logo of the heurigen and wine grower, Wieninger

Other examples include the logo of *Mayer am Pfarrplatz,* showing the profile of the old wine tavern, placed above the name of the *heurigen* written in a rectangle makes an allusion to traditional Viennese street signs.

Edlmoser uses a simple, yet modern writing style for his name, which is followed by the word 'Wien' or Vienna. He and his family operate his self-proclaimed "innovative" heurigen wine tavern, where he strives to emphasize both tradition and modernity.

Logo of the heurigen and wine grower, Mayer am Pfarrplatz

Logo of the winery and Heurigen, Edlmoser

RESULTS AND BEST PRACTICE IMPLICATIONS

The resurrection of the heurigen concept in Austria can be viewed as a best practice because it describes a successful change management process implemented by a variety of stakeholders. Even though today there are fewer active heurigen wine taverns in Vienna than there were a century ago, the ones that remain celebrate both the tradition of Austrian wine as well as attention to consumer needs. Because of this, they not only managed to survive in spite of all the changes in food and wine consumption in the 21st century, but also grew and established prosperous businesses with higher profit margins.

There are several reasons for their success, with the first being the *will to change* and the insight that an ongoing change process is crucial even in traditional offerings and environments. A component of best practice organizations is the ability to realize the need for change and move forward to implement a new vision (Blais, 2011).

Furthermore, many of the heurigens that survived were able to *overcome resistance to the change* that came from many sources: regular heurigen customers, winery clients and suppliers, their employees, the media, and often even their families. They did this by focusing on a vision of the future, working collaboratively, and adopting new marketing techniques.

Innovation was also another important component of the positive results. The ancient *Gemischter Satz* wine for example, had been forgotten in time, until it was rediscovered in a search to create new wine styles. A renewed focus on the environment and adopting organic and biodynamic processes is another form of innovation, in addition to the development of lighter wine styles to match consumer needs. The unique collaboration of the Wien Wein Group also illustrates innovative thinking by working together to promote the heurigen concept rather than competing against one another.

Partnerships with other agencies, such as the Vienna Tourist Board, the Austrian Wine Marketing Board, and the university researchers, were also a crucial part of the success for the heurigens who managed the successful transition into profitability. Without the collective efforts of all parties involved, especially the marketing and communication components, it would have been difficult for the heurigens to have as much impact as they did. Though there were no changes in legal regulations regarding restrictions on heurigen operation, the other partnerships did pay off and continue to do so.

In the end the heurigen is a unique business concept that defines a piece of Austrian history and culture, as well as serving as a distinctive method of wine distribution. Furthermore, it can be a very attractive and highly profitable business for those who are able to face and respond to the challenges resulting from new ways of consumption.

FUTURE ISSUES

Despite the success to date in saving the heurigens of Austria, there are still several looming issues for the future. These included the rising cost of labor in Austria that makes labor-intensive businesses such as wine production and running a heurigen more costly year after year. This development, unfortunately, has a second impact on talent management and succession planning. Many young talented professionals no longer want to work in wineries or heurigen because of lower pay, and the lure of more exciting modern jobs in other industries. Already some winery/heurigen owners are encountering succession problems in that their children are not interested in taking over the business.

In addition, a great number of new laws and regulations lead to an over-regulation of food offering businesses. Labeling allergens in restaurant or bar menus is, for example, a new European law. Ongoing discussions about the traceability of ingredients in meals and a possible new packaging law also threaten the industry and make it nearly impossible for small and regional suppliers to keep up with the new regulations.

In Europe and especially in Austria new anti-smoking laws and alcohol restrictions, such as lowering the allowed blood alcohol level from 0.8 to 0.5 per mil, led to significant losses in revenue in the catering industry. Furthermore, the rising percentage of Muslims in Vienna who do not consume the two key offerings of heurigen taverns, wine and pork products, is also of concern.

DISCUSSION QUESTIONS

1. Describe the different types of change in a wine tourism context and discuss which factors might differ in new versus old style offerings.
2. Identify reasons for traditional heurigen customers, employees and family members to resist changing and modernizing traditional heurigens. Then identify methods to overcome the resistance.
3. Identify other businesses that face similar challenges in a wine tourism context. Discuss ways in which they can change to be successful in the future.
4. What other steps do the heurigens of Austria need to take in order to be successful in the future?

Chapter Three

Winter Wine Festivals in Niagara, Canada

Linda Bramble and Carman Cullen
Brock University, Ontario, Canada

"As the temperatures begin to drop in Wine Country Ontario," said Magdalena Kaiser, "Winemakers prepare for the Icewine harvest while they watch in anticipation of the necessary -8C required for picking. It's an exciting time and visitors from all over the world have begun to anticipate the start of Niagara's Icewine Festival."

Magdalena is well aware of the many tourists who come to the Niagara Wine Region each year to participate in the Icewine Festival, because she is Executive Director with the Wines of Ontario organization. However, twenty years ago, there were very few wine tourists who dared to visit during the harsh Canadian winters. This was because as soon as post-harvest temperatures began to fall, the wineries of Niagara would shut their doors and wait until spring to welcome visitors again.

In the past, a Canadian wine tour in winter was not part of most winter travelling itineraries. International off-peak wine tourists were non-existent, and even hardy Canadians, ostensibly used to harsh winter weather, gave meaning to the term "getaway;" more likely to Florida, Mexico or Aruba, and not Niagara. However, by January 2015, things had changed. Tens of thousands of people arrived to enjoy three weekends of events at the Niagara Icewine Festival, named among the top most interesting festivals in the world by *National Geographic Traveler* (2014). What accounted for this innovative turn around in off-season wine tourism?

This chapter explores what winter wine tourism was like in 1995, prior to the first winter event held in 1996, and then examines how it grew in volume, economic return, and customer satisfaction compared with the high season of June-September. "When fate gives you lemons, make lemonade" is a hoary aphorism – but one that is applicable to the Niagara wine industry. Niagara winters can be brutal, with daytime high temperatures struggling to reach the freezing point – and often failing. But rather than complain, the wine business community marshalled their resources and energy to make something wonderful out of the hostile environment. This is the story of the last twenty years and the creation of winter wine tourism in the Niagara region of the province of Ontario in Canada.

OVERVIEW OF WINE IN CANADA

Maple syrup, Mounties, hockey, and bone-chilling temperatures all come to mind when one thinks of Canada – but grapes and wine? Canada is the world's second largest country by area, but very little of that land lends itself to growing grapes.

Nevertheless, fine wine is produced in six of Canada's ten provinces (British Columbia, Ontario, Quebec, Nova Scotia, New Brunswick, and Prince Edward Island); admittedly in small amounts in small pockets of land. It is interesting to compare the latitudes of grape growing regions around the world, comparing the Niagara region of Canada to other famous grape growing locales in France and the United States. In fact, eighty percent of the wine grown in the country is produced in the Niagara Peninsula of the province of Ontario, an appellation located on a strip of land between the Great Lakes of Ontario and Erie.

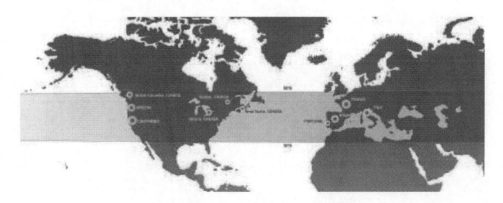

Map Showing Canadian Wine Region Latitude

There are roughly 500 grape wineries in Canada, and over 1700 grape growers/vineyards on more than 27,500 acres of land (CVA, 2014). The

Canadian appellation system is called the Vintner's Quality Alliance, or VQA. It is a similar quality designation to the AOC in France or DOC in Italy, ensuring that no imported grapes or juice are part of the product.

Canadian wineries producing 100% domestic grape wines are typically small enterprises having a capacity of under 500,000 liters. Approximately five percent of the Canadian wineries would be considered to be medium sized (500,000 liters to 1 million liters), and only one percent of Canadian wineries would be considered to be large, with annual production of over 1 million liters (CVA, 2014).

VQA sales in the province of Ontario represent 11% of total Ontario wines sales, with 30% share taken by Imported/Canadian blends and the remaining 59% totally attributable to imported wine sales (WGAO, 2013). Thus, unlike most wine growing regions of the world, far less than half of the wine sold in Canada has any Canadian content, and of the 41% that does, only 27% (or 11% of total Ontario wine sales) is VQA (purely Ontario wine).

Thirty-one thousand Canadians are employed in the wine industry and wine-related sectors (Canadian Wine and Grape Fact Sheet, 2013). The Canadian wine industry contributes $6.8 billion to the Canadian economy, and for every bottle of Canadian wine sold, approximately $31 of economic impact is generated (Canadian Wine and Grape Fact Sheet, 2013). Wine tourism currently adds $1.2 billion to the Canadian economy annually, with $644 million in tourism-related economic impact in the province of Ontario alone (Canadian Wine and Grape Fact Sheet, 2013).

OVERVIEW OF WINE IN THE NIAGARA PENINSULA

The Niagara Peninsula is located in the Canadian Province of Ontario in the eastern part of the country. It borders New York State and is nestled along Lake Ontario and the famous Niagara Falls. Legislated as a wine appellation region, it contains several sub-appellations and 83 wineries (VQA Ontario, 2014).

# of Wineries in Canada	500
# of Wineries in Niagara Peninsula	83

Roughly the size of Napa Valley, the Niagara appellation has 54,000 acres of arable land with 13,500 of those acres currently planted to *Vitus*

Vinifera varieties (VQA Ontario, 2014). The area has been producing wine since the mid-19th century, originally based on the native varieties of the labrusca species, because the received wisdom of the time was that classic European varietals could not be grown in the "frozen North." The labrusca grapes and their hybrids were hardy, but produced wines of varying quality that were sometimes made more potable by fortification or by adding international wines to local cuvees.

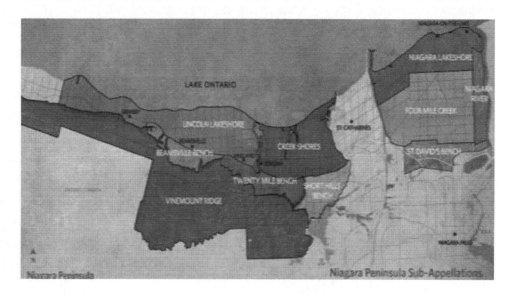

The Sub-Appellations of the Niagara Peninsula Appellation

In the early 1970's it became apparent to some pioneering growers and entrepreneurs that the more cold-sensitive *Vitus Vinifera* varieties could and should be grown in the region. A modern industry began to grow specializing in such cool climate varieties as Riesling, Chardonnay, Pinot Noir, and Cabernet Franc, and, with surprising success, Cabernet, Sauvignon, and Merlot.

Re-branding the industry to differentiate it from the stigma of the past has been a tough row to hoe among those wine drinkers who remembered the old industry, many of whom still cannot believe that quality wine can be produced in Niagara. Thankfully, enlightened drinkers and newer generations are more open to experiencing wines made in Niagara, many of which have garnered some of the most prestigious wine awards in the world. It is a small appellation on a world scale, but capable of producing age-worthy premium wines.

In the summer months, beginning in the late 1970's, the transformation in product quality and increasing consumer acceptance of the product started to bring thousands of visitors to the highly capitalized properties in Niagara, and for that, the industry was very grateful. The hordes of tourists visiting Niagara Falls each summer now had an additional outing to add to their itinerary; after all, one can only stand and watch water cascading over a cliff for so long. The wine tourism industry blossomed and became a major contributor to the overall tourist experience in Niagara.

THE PROBLEM: LACK OF TOURISTS DURING THE COLD WINTER MONTHS

As wine tourism continued to grow each year during the warm summer and autumn months, some local vintners, hoteliers, and restaurateurs began to ask the question of how they could continue to encourage the stream of visitors during the winter months. Though many wine regions around the world struggle with this issue, the Niagara region of Canada has a special wine that thrives in winter – the Icewine. In fact, it is required by law that the temperature fall as low as -8 Celsius before the frozen grapes with their high sugar content can be harvested. How could the industry capitalize on this unique aspect of the Canadian winter?

THE SOLUTION: HOW NIAGARA CREATED A WINTER WINE WONDERLAND WITH THE ICEWINE FESTIVAL

The story of Niagara's winter wine tourism begins in 1995. The new owners of Hillebrand Estates, Andrew Peller, Inc., invested several millions of dollars on fixed assets making this mid-sized property tourism-ready. That meant providing better access to the winery for visitors, expanding the tasting room, improving the restaurant, and expanding the cellar. To have all of that investment languish in the winter months was not part of their business plan. True, visitors came in December, but January to March was a dead zone. Most wineries simply closed their doors until spring.

The Hillebrand Estates challenge, according to Greg Berti, General Manager at the time, was to create something meaningful to attract people to visit the winery in winter. "Doing what we normally did in terms of tours and tastings," he stated, "was not going to be good enough."

Therefore, Berti gathered his colleagues together and said, "Let's take this head on! What do we do best in January?" The answer came in a common chorus of replies-Icewine!

Icewine has become an iconic product for Niagara. Visionaries such as Karl Kaiser and Don Ziraldo are credited with borrowing a technique from Germany, and transforming an industry. Winter wine tourism would not exist without this elixir. Made from rock-hard frozen grapes picked between -8 and – 12 degrees Celsius, it is a very expensive product to make and buy.

During January of most years, the grapes for Icewine are harvested, so celebrating Icewine was a natural theme. The possibilities seemed endless. That vision of the possibilities inspired them. "Painting a picture of what something could be," said Berti, "was a critical element of the project."

The challenge was to time the event late enough in January so people were over their December spending remorse, but still early enough that the vineyard actually had icewine grapes on the vine. Part of the visitor experience, the planners reasoned, would be to pick the snow-covered, frozen grapes, and then see them pressed. That signature extraordinary winter experience became iconic, like no other in the New World.

Frozen grapes waiting for -8 degrees Celsius to be harvested.

Berti recalled his early fascination with the visitors who attended their little 'festival'. "We'd leave two or three rows of icewine grapes on the vine and people wouldn't stop picking. We wanted to save some for the next day, but they just kept on picking, dutifully carrying their bushels of grapes in to watch them being pressed."

The idea was to engage the bundled up visitor who had to move around to keep warm, so the planners at Hillebrand added other activities to keep visitors entertained. These included roasting chestnuts, serving hot soup outdoors, and providing horse-drawn carriages around the vineyard. They also organized chocolate-making demonstrations, offered live music, and provided specials tasting of current and back vintages of Icewine. One of the most successful events was an ice sculpture competition involving up to twenty carvers sculpting up and down the Hillebrand walkway with a $1000 prize for the winning sculpture.

"We did all those things," said Berti, "to make it festive in nature." At the winery restaurant they created five to seven-course dishes pairing it with Icewine or using it in the recipes.

Ice Carving with Icewine and Grapes

Hillebrand Estates held the event on their own for the first five years, but they felt that it wasn't going to be a meaningful festival unless more wineries were involved. Therefore, they invited other wineries to participate, and this became the tipping point of visitor transformation. With more involvement, creative ideas developed such as offering a "passport" to each of the participating wineries, which included a tasting and another specialty attraction depending on what each winery wanted to offer.

One of the smartest policies the organizers set in the early stages of the festival's evolution was to avoid planning a winter festival dependent on winter activities such as cross country skiing or ice skating. Though counterintuitive, it was imperative to the event's success.

"Niagara can have some very warm Januarys," explained Debi Pratt, a public relations manager for Jackson Triggs and Inniskillin wineries. "Our planning idea was not to design it around the weather. Rather, we planned enough activities that would make the event successful without depending on the weather being cold."

There were resistors. The smaller wineries were tired. They had just finished harvest and couldn't justify bringing in staff on a weekend, so Hillebrand, as a larger winery, funded some of the smaller players just to get them on board. Once the smaller producers saw the benefit in terms of sales and exposure, they became continuing participants. Sherri Lockwood, Marketing Manager for Hillebrand Wineries, explained the situation. "Smaller wineries needed evidence that there was activity enough to make their cash registers ring. Short-term wins were critical in the early days."

The next factor that became fundamental to the festival's success was engaging other partners such as the hotels and inns, restaurants and retail shops. They needed to work together creating weekend packages in conjunction with the festival. Fortunately, the Chamber of Commerce, under the leadership of Janice Thompson, stepped in to assist with the festival. She reached out to multiple stakeholders and additional wineries to gain commitment and new ideas.

"What had to occur for this event to be successful," reported Thompson, "was the input and cooperation from the wineries as to what they wanted the event to look like. For instance, they wanted a unified décor with beautiful banners printed, keeping things natural and clean."

Thompson positioned the event as starting from scratch with the wineries and asked, "What can we do that is distinctly 'Niagara on the Lake'?" Together they developed some unusual activities such as a sparkle and ice event that kicked off the Icewine village on the first Friday. Held inside their historic courthouse, they invited seventeen restaurants to participate by pairing signature culinary dishes with local wines.

At 10:30 in the evening, under the glow of thousands of twinkling lights, each visitor was given a glass made of ice, with a little bag that contained a cubic zirconia with a number. The person with the winning number won a real $4800 diamond. For the daytime events on Saturday and

Sunday, they constructed a thirty-foot tasting bar made of ice with twenty-eight wineries participating.

Today, the Icewine Festival continues to innovate, and every stakeholder has a role to play. The larger wineries provide the signage with posts and arrows. Ice sculptures are still part of the event. A new and successful addition was the introduction of an Icewine Cocktail competition. The Chamber of Commerce continues to participate by managing the administrative aspects of the festival. This includes overseeing wine inventory, rules, permits, safety of visitors, insurance, and security. While in the beginning marketing was accomplished by individual wineries and word-of-mouth, today it includes the participation of *Wine Country Canada* who assists with developing brochures, the website, ads, and obtaining media sponsorships.

New leaders have also stepped forward to help, such as Andrea Kaiser, whose father was the co-founder of Inniskillin – one of the most famous producers of Canadian Icewine. Andrea brought the different interest groups together to agree on such things as messaging and establishing a common brand. She also helps with fundraising, because the festival can be expensive to executive, with the ice bar alone costing $10,000 to construct each year. "Working with nonprofit organizations has been limiting," explained Kaiser. "We are all volunteers with full-time jobs of our own, but we're passionate and want to make it work."

The Twenty Valley Winter WineFest

The Niagara Icewine Festival became so popular that a neighboring wine region, the Twenty Valley near the town of Jordan, decided to join them with a slightly different version of the Icewine Festival. Similar to the event in Niagara on the Lake, they installed an ice bar with wine samples, buskers, street performers, and ice sculptures.

Two main features differentiated the Twenty Valley event from the one held in Niagara on the Lake. The first was its duration: in Jordan the event is held on one weekend only; Niagara on the Lake is held over three weekends. The second distinction is the fact that they serve table wines along with Icewines. Many of their participating producers do not make Icewine, but do specialize in sparkling wines and red and white table ones.

J.D. Pachereva, Executive Director for the Twenty Valley Winter WineFest, described the success of the event. "The first year we ran the event we had thirty wineries participating and the traffic we experienced was beyond our expectations. We ran out of glasses. Having never done the event, I thought we were safe with 2,500 glasses, but we ran out of them by 2:00 on Saturday. I ran around and found every plastic glass I could find. I remember thinking, in the midst of my anxiety, 'I think we're on the right path.'"

Wine Barrel rolling competition during Winter WineFest

Since then they have integrated outdoor fire pits, an Icewine dinner with a celebrity chef, and a brunch. In addition, they feature a fashion show, live music, a winemaker barrel rolling challenge, and an after-party at Cave Spring Cellars, one of the Canada's most acclaimed wineries. Along with serving a diversified list of wine styles and a culinary component with local chefs, they have local and national Canadian entertainment. To help defray costs they applied for, and received, a provincial tourism grant from "Celebrate Ontario," which has contributed for the past four years.

RESULTS AND BEST PRACTICE IMPLICATIONS

The Niagara Icewine Festival is considered to be a global best practice in wine tourism for several reasons. One is primarily due to its distinctiveness as one of the most unique wine tourism events in the world, and has been named as such by National Geographic Traveler (2014).

Furthermore, it illustrates how working together, the Canadian wineries and other key stakeholders in this region, literally transformed the cold dark winter months into a sparkling fairyland of fun and wine for tourists. In the past, when no visitors came at this time of the year, now more than 15,000 tourists flood the region for three weeks to taste Icewine and celebrate the winter harvest.

Because of this, wine tourism revenues have blossomed. Today the province of Ontario welcomes 1.9 million visitors each year and contributes $3.3 billion to the Canadian economy; fully half of the entire Canadian Wine and Grape industry (CVA, 2014). Of this, the single largest contributor to the Province's wine economy is the Niagara region. There are several implications that can be gleaned by examining the factors that contributed to the success of the Icewine Festival. These are summarized in the following paragraphs.

The Power of Vision and Focus

The festival started small and grew incrementally. Participants reported that it was very important to keep their eyes focused, with forbearance, on the vision they had conceived and chose to follow. If things went sideways they redrew the tactic, not the ultimate direction in which they were headed. As Greg Berti said, "Painting a picture of what something can be was a critical element of the project."

The Power of Partnerships and Collaboration

A large amount of collaboration with different partners was an important ingredient to making the festival a success. Without the willing participation of stakeholders such as the wineries, local hotels, restaurants, and bed and breakfasts who contributed by developing winter packages the festival could not happen. The other aspect of partnerships was the necessary commitment of the key players ---the wineries who were committed to 'making it happen,' the tourism associations who assumed the administrative responsibilities and the larger wineries who were willing to support the early stages with investment of time, money, and personnel.

The Power of Innovation

Although the planners on both sides of the Niagara Peninsula maintained the defining elements of the festival such as the outdoor ice bar, culinary events and outdoor ice sculptures, it has been the addition of new and innovative venues and activities that has kept the festival fresh. By adding something novel every year, such as the Icewine cocktail competition, or the glasses made of ice, this keeps visitors returning again and again to see what's new.

The Power of Leadership and Culture

Winter wine tourism in Niagara would not be the success that it is today without the leadership of Greg Berti and the other visionaries who put in the effort and resources to build a unique, iconic addition to life in Canada. Winters in Canada can be bleak and inhospitable, but the players in the wine industry shortened it by bringing light and life to Niagara with the Icewine Festival. Even more important is that the festival highlights the true character of Canadians – that of hospitable, hard-working, fun-loving people who embrace winter.

FUTURE ISSUES

The Niagara Icewine Festival is one of the few events in the Niagara Peninsula that started as a grass roots initiative and continues to be so today. Other larger stakeholders have joined them such as Wine Country Canada and the Province of Ontario, however, it remains a grass roots, locally conceived and operated festival. One operational imperative will be to make sure that there are sufficient accommodations for guests. In the Niagara on the Lake side of the Niagara Peninsula there are a plethora of accommodations, but this is not the case in the Twenty Valley side. This could present an issue for them in the future if the festival continues to expand in size.

Global warming could also be a long term future issue for the Niagara region. According to some experts, by the year 2050 it may be too warm to produce Icewine in this part of Canada (Goldenberg, 2013). This would mean that the Niagara Icewine Festival would need to either move further north, or adapt by adding other wine styles.

There is a possibility that over time the organizers of the event could lose focus or become less collaborative. This could harm the innovative spirit and energy of the festival. However, if they continue to volunteer their time and energy to lead, attend meetings, and listen to different points of view, the festival organizers will continue to grow the event in a positive fashion.

Finally, just as there is a terroir in winemaking, there is also a "Terroir of Tourism" that cannot be ignored. As long as the Niagara Icewine Festival continues to reflect the local reality of what the winter wine tourist demands: a unique, Canadian winter experience, then the authenticity of the event can be sustained. Authenticity is the *sine qua non* of wine tourism, as it is for winemaking.

DISCUSSION QUESTIONS

1. What are the primary learning points that are generalizable to other wine regions around the world?

2. Do you think that off-peak winery tourists are different than "regular" winery tourists in any managerially relevant fashion? Discuss.

3. Greg Berti was amazed that people would pay money to leave the comfort of their homes, in the middle of winter, generally at night, to pick frozen grapes that would be made into Icewine. Does this amaze you? What explains this kind of behavior?

Chapter Four

Building a Wine Destination from Scratch in Ningxia, China

Wenxiao Zhang & Liz Thach

Grape & Floriculture Bureau, Ningxia, China & Sonoma State University, California, USA

The highest peak of the Helan Mountains rose over 3500 meters, towering over the irrigated plain that was home to the more than 39,000 vineyard hectares in the Ningxia wine region, located in Central China. Glancing up, Mr. Kailong Cao, Director of the Bureau of Grape and Floriculture Development for the province, could see a dusting of white snow on the far peaks on this sunny December morning.

In front of him, moving slowly across the sandy plain that was blessed with the alluvial soil washed down from the Helan Mountains, were six large bulldozers creating more vineyards for the burgeoning Ningxia wine region. Mr. Cao smiled as he realized his dreams of creating a world-class wine tourism destination were coming true. He remembered twenty years ago when there was nothing here but vacant land. Now this valley, situated at 38 degrees north latitude, the same as Napa Valley, was creating some of China's best wines including *Jai Bei Lan*, which won the *Decanter World Wine Awards* in 2011.

Mr. Cao turned to address the group of international journalists standing behind him snapping photographs and scribbling in their note pads. His breath was visible in the frosty air as he described in Chinese all of the progress they had made in the last several years. His translator expertly relayed his message in English and French, and then the visitors began to ask questions. Mr. Cao smiled, delighted to describe the journey that had allowed him and his team to create a wine destination in Central China from scratch.

OVERVIEW OF THE CHINESE WINE INDUSTRY

China has been producing wine since 1600 BC, according to scientists who discovered ancient pottery vessels in the Henan Province, with organic residue revealing a fermented beverage of rice, honey and fruit (China Culture, 2003). However, the term "wine" in Chinese is "*jiu*," and refers to any type of alcoholic beverage. It is important to clarify grape wine by using the term "*putaojui*." For centuries, rice wine has been much more popular in China than wine made from grapes, though China does have its own indigenous grape varieties - *vitis amurensis.*

Wine produced from *vitis vinifera* grape varieties was first made in China in the 1890's in the town of Yantai in the Shandong Province. It was here that Zhang Bishi, a Chinese businessman and diplomat who traveled frequently overseas, decided to open the first large-scale winery and called it Changyu Pioneer Wine Company (Lyons, 2013). Importing wine grape cuttings from Europe, he found that many did not survive, so he grafted them to the native Chinese rootstock and had much more success. Eventually, he began winning medals for his wines at international competitions. Today Changyu is one of the largest wineries in China, with eight massive chateaux in various regions of the country, as well as international holdings (Changyu.com, 2014).

Eventually, during the mid to late 1900's other grape wine establishments began to appear in China, including the giant COFCO winery operated by the Chinese government, which produces the famous *Great Wall* wine brand. According to the StatOIV (2014), China's grape vineyards grew from 300,000 hectares in 2000 to 560,000 hectares by 2011, and their world ranking was fifth largest in wine production in 2012 (OIV, 2014). The latest report on the number of wineries in China was 625 (Gastin, 2014).

Major Wine Varietals in China

According to Jiang Lu (2015), a professor at the China Agriculture University in Beijing, 80% of the grapes grown in China are table grapes, with 15% used as wine grapes and 5% for raisins. He cites that major grape varietals planted in China include Cabernet Sauvignon at 49.6%, Carmenere at 9.6% (also referred to as Chinese Gernischt), Merlot at 8.5%, and Syrah and Chardonnay, both at less than 2%. More red wine is sold in China than white wine, primarily because the color red is thought to bring good luck, but some sources report that white wine in gaining in popularity in China (Lawrence, 2012).

Eight Major Wine Regions of China

Currently there are eight major wine regions in China (Johnson & Robinson, 2013):

1) **Shandong** – the oldest wine-producing region with the largest number of wineries. Located on a peninsula that juts out into the Yellow Sea, it has a more moderate climate, but often has high humidity that causes mildew problems for wine grape production.

2) **Xinjiang** – the second largest grape producing region, but with a large percentage of table grapes as well as some wine grapes. Located to the far west of China, the region is dry and arid, but has very cold winters that require the vines to be buried underground to protect them. A primarily Muslim area of China, it is part of the original Silk Trail and has towns such as Turpan where they have been producing wine since 300 B.C. (Thach, 2009).

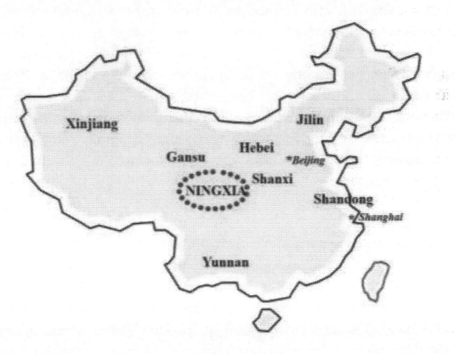

Map of Major Chinese Wine Regions

3) **Ningxia** – located in central China near the Helan Mountain range, Ningxia is known for its dry, sunny climate and production of high quality wines. Though it is still often cold enough to require the vines be buried in the winter, the climate is friendlier to producing healthy grapes, and the alluvial soils and Yellow River flowing nearby provide positive conditions for wine grapes.

4) **Hebei** – located just outside the major city of Beijing, Hebei is home to some very large wineries. It has the added advantage of luring many tourists due to its proximity to the Great Wall of China and the Summer Palace.

5) **Shanxi** – located west of Beijing, Shanxi is famous for the headquarters of Grace Winery, which was one of the first wineries in China to achieve acclaim for its high quality wines produced with *vitis vinifera* grapes.

6) **Gansu** – also located in central China, Gansu is a newer viticulture area in a cooler location north of Ningxia.

7) **Jilin** – located to the far northeast of Beijing, Jilin is a much cooler area renowned for its ice wines.

8) **Yunnan** – also referred to as Hunnan, is the most southern wine region of China close to the borders of Laos and Burma. Though a warmer region, most of the vineyards and wineries are located in the mountains at altitudes of up to 9800 feet (Johnson & Robinson, 2013).

There are some other regions in China also growing wine grapes, such as Shaanxi (not to be confused with Shanxi) and Tianjin near Beijing, as well as a few emerging areas in others parts of the country.

Wine Tourism in China

According to the World Tourism Organization (UNWTO, 2015), China ranked first in most visited countries in North Asia with 136 million tourists visiting in 2014. Interestingly China was also the world's top tourism source, in that many Chinese citizens visited multiple countries around the world to spend a total of $165 billion on tourist activities outside of China in 2014 (UNWTO, 2015).

In terms of wine tourism, China is still in its infancy regarding international tourists visiting Chinese wine regions. Part of this has to do with the fact that the industry is focused more on attracting national Chinese tourists than non-nationals (Cao, 2014). Though there are a few tour companies that advertise wine tours, the fact that international visitors are still required to obtain a visa, and often an invitation letter, makes this more challenging. Despite this fact, thousands of Chinese tourists visit the amazing wine chateaux and tasting rooms that have been constructed in some of the major Chinese wine regions. Some look like European castles, whereas others use traditional Chinese fortress architecture.

OVERVIEW OF THE NINGXIA WINE REGION

Ningxia is one of 34 provincial level administrative regions in China, and has its own government that reports to the central Chinese government in Beijing. Located in Central China, Ningxia vineyards lie in a protected valley and are sheltered from the cold Northern winds by the Helan Mountains, that rise over 11,000 feet tall at their highest peak. The soil consists primarily of sand and rocks, with many alluvial fans flowing down the sides of the mountain to form an ideal grape growing foundation, similar to the alluvial fans found in Napa Valley. The Yellow River flows near the vineyards and through the nearby capital city of Yinchuan. Elevation of the vineyards averages 3,000 feet, with most vines established on the flatter plains of the alluvial fans, but a few planted on the foothills of the Helan Mountains (Cao, 2014).

# of Wineries in China	625
# of Wineries in Ningxia	72

Situated at 38 degrees north latitude, the climate is Continental with average summer highs of 74 F, though it can easily reach into the 90's F. Winters can be harsh with some snow and average temperatures of 10 F. Rainfall is low, ranging from 8 to 27 inches per year, but almost all vineyards have drip irrigation. The trellis system is primarily vertical shoot positioning (VSP) with cane pruning. The trunks of the vines are not allowed to grow very wide, because they must be buried during the cold winters. Currently more than 50% of the vines are buried mechanically (Cao, 2014).

The first vineyards were planted in 1982, and by the end of 2014 Ningxia government officials reported that they had planted roughly 39,000 hectares of vines, and have 72 wineries in operation (Cao, 2015). Expansion and promotion of Ningxia as a wine region has been encouraged by the Chinese central government, and funded by the Ningxia government (Cao, 2015). Currently, they are engaged in an ongoing effort to attract private developers and international wineries to establish new vineyards and wineries in the region.

Interestingly, the term for a Chinese acre is mu, and equals $1/15^{th}$ of a hectare. Therefore 39,000 hectares would be roughly equivalent to 585,000 mu. Since a hectare equals 2.47 acres, this would be equivalent to 96,330 acres.

Winter Vineyard in Ningxia

As Ningxia is a relatively new wine region, growers are still experimenting to determine the best type of varietals to cultivate in their climate. Therefore, they have currently planted 40 varieties, but the main ones are:

- *Cabernet Sauvignon* – the cabernets from this region are medium-bodied with a light ruby non-opaque color. They are fruit forward with concentrated red and black berry flavors and spices. Ningxia wineries have won the most awards on their cabernet sauvignon and cab blends.

- *Merlot* – often darker in color than the cabs, with savory meaty flavors and some herbal components.
- *Chinese Gernischt* (also called Chinese Cabernet) - this varietal has been proven to be the same as Carmenere (Robinson, 2012). In China, it seems to produce wines that have a strong herbal, almost green note with dark, smoky fruit and astringent tannins.
- *Italian Riesling* – floral, fruity with some tropical notes of pineapple producing light and approachable wines with a crisp acidity.
- Other grape varietals include: chardonnay, pinot noir, cabernet franc, and vidal blanc; the latter of which is used to make ice wines.

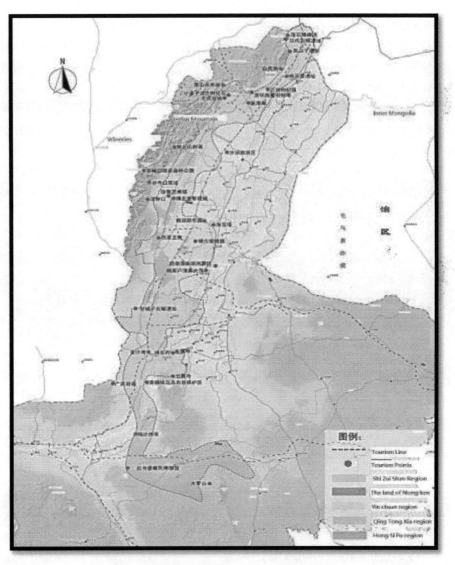

Map of the Ningxia Wine Region

Tourism in Ningxia

The capital of Ningxia is Yinchuan, a city of around 2 million people. It is located only 2 hours by plane from both Beijing and Shanghai, and receives thousands of Chinese tourists each year who come to visit some of its famous tourist sites. These include the Xia Tombs, Sand Lake, Helanshan Rock Paintings, the Drum Tower, the 108 Dagobas, and several other museums and temples. Yinchuan is also famous for the Automobile and Motorcycle Tourism Festival that is held every July.

In terms of wine tourism, the Ningxia government has not yet started to actively promote visits to the wineries for international tourists. Domestically, Ningxia is using more of a "soft launch approach," with no strong promotion of wine tourism at this time. Yet domestic tourists who arrive in Yinchuan have the opportunity to visit some of the chateaux if they make appointments in advance. They may also be able to attend the Ningxia Grape & Wine Festival, which includes a large trade show with wine suppliers, wine tastings, excellent food, and entertainment such as dancers and singers.

Ningxia Lamb Dish with Wine

Jiabeilan Wine from Chateau Helan Qingxue

The Ningxia region is also known for its unique cuisine, including tender young lamb cooked in broth, organic vegetables, and fresh fruit in season. Another specialty of the region is the Chinese wolfberry, which is a small red berry, usually served dry and added to soups and salads. It has a taste that is similar to dried tomatoes. According to China Today, "Ningxia is known for five specialties or treasures, described as "red, yellow, white, black, and blue" because of their respective colors." These are the "wolfberry, sheep fur, liquorice root, facai, and Helan stone (2014, p. 1)." Facai is a moss that is also known as the "hair vegetable," because it has a soft texture like vermicelli and the appearance of black hair (Clove Garden, n.d.). The special stone of the Helan Mountains is a beautiful polished granite, which is often dark green and/or purple in color, and is one of three famous ink stones in China that are water resistant (China.org, 2011).

THE PROBLEM: NO GLOBAL RECOGNITION AS A FINE WINE GROWING REGION

Even though the first vineyards were planted in Ningxia in 1982 by Yu Quanying Farm as a means to provide employment opportunities for local workers and to expand the agriculture business of the region, it wasn't until the late 2000's that the concept of Ningxia as a world class wine region came into being. At that time, Mr. Cao and his staff at the Bureau of Grape and Floriculture Development were trying to identify strategic growth businesses for the region. After conducting some studies they realized that the terroir was ideally suited to wine grapes. Mr. Cao reflected on this discovery:

> *"We realized we were located in a "golden zone" for vineyards in China. Our studies showed that the east foothill of Helan Mountain has all the natural conditions needed for premier wine grape cultivation. This includes adequate sunlight and favorable temperature range, pollution-free soil, decent rainfall and irrigation. Even more important, we realized that the environment produces aromatic grapes with balanced sugar and acid levels."*

The problem, however, was how to gain recognition that Ningxia could be a world-class wine region. At the time, most people didn't think that China could ever produce high quality wine. For the most part, this perception was justified in the early 2000's, because much of the wine that was produced in China was bulk wine with no clear regulations on additives or quality. Also, because the Chinese population had a long history with rice wine and high alcohol *baijiu*, most people didn't know how good wine was supposed to taste.

To complicate the problem further, international wine experts had never heard of the Ningxia wine region, so there was no global recognition. Other issues had to do with the need to develop world-class winemaking skills, obtain more financial backing, plant more vineyards, build wineries, and develop a wine tourism infrastructure. Taken altogether, Ningxia had the daunting task of trying to develop a world-class wine destination from scratch.

THE SOLUTION: DEVELOP STRATEGY TO BECOME A HIGH QUALITY GLOBAL WINE REGION TO ATTRACT TOURISTS

Fortunately, the timing was good for Mr. Cao and his staff, because the Chinese central government in Beijing was encouraging new agriculture projects, such as wine grape production, for several reasons. The first was that they preferred to use rice as a food product instead of a means to produce rice wine, and secondly they were concerned about the high alcohol content of *baijiu*, often 40 to 60 proof, and the potential negative impact it had on health. Additionally, wine grapes needed less water than many other crops and could be grown in places where there was currently no agriculture. Therefore viticulture and wine production was viewed positively at the time. Indeed the Chinese government invested in the Great Wall wine brand as part of their state run China National Cereals, Oils and Foodstuffs Corporation, or COFCO.

With the blessings of the central government, the Ningxia government decided to fund the development of a world-class wine region, with the expectation that private funding would also be solicited in the future. In order to do this, they implemented the following steps.

Step One: Develop a Vision & Strategy

Strategy experts have frequently pointed to China as being an expert at visionary thinking and long-term strategy (The Economist, 2014), and in the case of the Ningxia wine region vision and strategy, this seems to hold true. In the mid 2000's the Ningxia government created a vision to:

> *Expand to more than 1,000 wineries in the next decade, and create a world-class wine tourism destination for Chinese tourists by 2016 and international tourists by 2017. The region will have a wine route with great chateaux, vineyards, and a chain of wine and cultural tourism experience focusing on Health, Sport and Relaxation. It will be renowned as the premier wine-growing region of China.*

The strategy to implement the vision included a three-pronged approach to 1) develop human resource talent to produce world-class wine; 2) execute a quality control system to ensure world-class wine; and 3) create a world-class wine tourism infrastructure and positive regional branding to promote it.

In order to accomplish this, the Ningxia government appointed the Bureau of Grape and Floriculture Development to manage the process and set up a regulatory system for grape and wine production. This made the Bureau the first provincial level institution in China to be charged with the management of a regional wine industry. It also highlighted Ningxia as the first region in China to create stringent regulations for local viticulture and wine quality production.

Step Two: Obtain Financial Backing

In the beginning the Ningxia government provided the initial funding to train employees in world-class viticulture techniques and wine production, as well as to provide money to bring in international experts to advise and evaluate. However, as Ningxia started to gain more positive press with articles and wine awards, both domestic and foreign enterprises, such as French conglomerates LVMH and Pernod Ricard, were encouraged to invest in the region (see step 5).

Step Three: Invite Global Wine Experts and Train Locals

During the early implementation of the strategy, the main focus was on the development of human resource talent. Therefore the Ningxia government encouraged interested locals in studying viticulture and winemaking in Bordeaux. They also hired many winemaking consultants from France, Australia, the USA, Italy, and other regions to provide advice on what they could do to improve viticulture and winemaking practices. They invited professors and other experts from around the world to make recommendations on how to implement a wine tourism infrastructure and build a regional brand.

The Ningxia government also partnered with local universities to establish educational programs to train locals in viticulture, winemaking, and wine tourism. They worked with Ningxia University, Ningxia Polytechnic School, the Ningxia Institute of Prevention and Control of Desertification, and helped to establish the Wine School of Ningxia University. In addition, the Ningxia government carried out a series of activities and educational events for local wine industry participants to communicate with their foreign counterparts in order to improve their skill levels.

Step Four: Establish Quality Metrics

Probably one of the most important steps Ningxia took was to establish quality control metrics for viticulture and wine. They began by encouraging best practices in viticulture, including careful rootstock and clone selection with no disease, proper vineyard set-up and installation, the use of drip irrigation, and encouragement to use organic and/or sustainable farming practices. They also established regulations on the amount of grapes that could be harvested from a high quality vineyard, as no more than 500kg per mu. This equates to 7.5 tons per hectare, or 3 tons per acre. They created a Geographical Indicator and named the region "Helan Mountain's East Foothill Wine Region", which is frequently shortened to "Ningxia Wine Region."

In terms of winemaking regulations, they established that at least 75% of the grapes in a bottle must be grown in the Ningxia region. In addition, 85% must be of the same variety and vintage in order to list varietal and vintage on the bottle. These standards are similar to those established by other major wine regions of the world, and sanctioned by the OIV.

Probably most impressive was the introduction of a winery classification system. Modeled on the Bordeaux 1855 Classification, but with updated rules and a required renewal every two years, it ranks wineries into five levels. In order to be considered, wineries must make at least 4,166 cases and farm at least 13 acres of vineyards (Thach, 2014).

An international group of wine experts including viticulture specialists and educators is brought in to evaluate the wineries, which Ningxia refers to as chateaux. Quality of vineyards, wine, and tourist attractions, including restaurants and lodging, are judged. During the first year of implementation in 2013, the judges selected 10 wineries as fifth-growths. Every two years wineries will be re-evaluated and eventually some will be promoted to fourth growth, third growth, etc. The long-term goal is that there will be chateaux classified in all five levels. Wineries can also be demoted or dropped from the classification if they do not maintain high enough quality levels (Thach, 2014).

Further, the Ningxia government established a quality control system for wine tourism. To do so they commissioned the International Center for Recreation and Tourism Research (CRTR) to evaluate the ranking of tourism wineries in Ningxia (Cao, 2015).

Step Five: Encourage Financial Investment

In order to implement their vision of a vast wine route, the Ningxia government invited domestic investors to establish new vineyards and great chateaux with tourist facilities, such as tasting rooms, restaurants, lodging, game rooms, spas, and other appealing venues. Fortunately they launched this phase of the strategy implementation when the economy was booming in China, and therefore many wealthy people invested in the project.

They also established conditions listed in the Regulations on Conservation of Helan Mountain's East Foothill Wine Region. These required that each investor plant a virus free vineyard of at least 13.3 hectares, and that it be in operation at least two years before they were allowed to build a chateau. Investors were strictly evaluated to ensure they were serious about creating wineries, and to prevent unqualified and speculative transactions-motivated developers from entering the region. Additionally, they set up regulations stating that any chateau not meeting the standard of the wine region would be removed. The minimum investment for a newly constructed chateau was set at RMB 20 million yuan (about USD 3.2 million). Naturally, in China, these types of strict regulations created a prestige buying opportunity, and spawned more desire on the part of the wealthy to buy into the project and build their own chateau.

Chateau Changyu Moser XV in Ningxia

Foreign investors were also encouraged to establish production facilities in Ningxia, and once word spread about the positive aspects of the terroir and awards on some of the early wines, this soon attracted others. The French were the first to arrive, with LVMH establishing a large sparkling production facility, and Pernod-Ricard following soon after. In addition, two of the largest and most famous wine brands of China set-up operations in Ningxia, namely Changyu and Great Wall.

Finally, the handful of regional wineries that had already existed within Ningxia were encouraged to expand and focus on high-quality production. Many were given educational opportunities to enhance their skills sets, and assistance with marketing and other types of support.

Step Six: Build a High Quality Global Brand to Attract Tourists

At the same time as Ningxia was engaged in establishing their quality metrics, they started a slow and subtle branding strategy. It began by inviting handpicked international journalists and sommeliers to the region to taste wines, visit wineries, provide feedback, and learn about the regional wine vision. These experts returned home and published articles, which created some positive press.

Ningxia also entered their wines in Chinese competitions for several years, learning about needed improvements from the judge's feedback. Eventually in 2011, they decided to submit wines into the *Decanter World Wine Awards* competition in London. They were just as surprised as the rest of the world when Chateau Helan Qingxue won the Red Bordeaux Varietal Over £10 International Trophy. Out of over 12,000 wines entered, only 25 international trophies are awarded at this competition (Lechmere, 2011). The wine was a 2009 Cabernet blend called Jiabeilan. It was made by two Chinese winemakers both trained in Bordeaux: male consulting winemaker, Li Demei, and female head winemaker, Zhang Jing.

The wine was a blend of cabernet sauvignon, merlot and cabernet gernicht (carmenere), and judges described it as "supple, graceful and ripe but not flashy…, excellent length and four-square tannins (Lechmere, 2011, p. 1)." This single event made headlines around the world, and immediately created even more positive press for the Ningxia wine region. More professional wine experts and journalists wanted to learn about the area, and Chinese nationals were thrilled that a Chinese wine had garnered such acclaim. Though 20,000 bottles had been produced, it nearly sold out overnight as many people in China clamored to purchase a bottle.

Then in 2012, the Ningxia government introduced another "marketing" scheme designed to bring global attention to the region. They established the "Winemaker's Challenge," in which winemakers from around the world were invited to apply to win a free trip to Ningxia to make wine during harvest (Robinson, 2014). This event received much international press, and helped put Ningxia on the world stage as a winemaking region that was serious about making good wine. Since that time, they have repeated the Winemaker's Challenge in 2015, and hope to do so again in the future.

Step Seven: Design a Wine Tourism Infrastructure

The Ningxia government invited experts in wine tourism to advise them on how to develop a wine tourism infrastructure. Over the course of several years, they created a wine route through the vineyards in the large valley that lay at the base of the Helan Mountains, and created a map showing where wineries were located. By the end of 2014, there were a total of 72 wineries in production (Cao, 2015), with many having tourist facilities such as tasting rooms, tours, restaurants, and lodging.

Because the wine region is located about an hour drive from the city of Yinchuan, they also focused on developing tourist support systems there as well. The airport was spruced up for visitors, and new welcome signs were erected. Tour companies with vans to carry visitors were encouraged to develop tours of the wine region for domestic tourists. In addition, they worked with the city of Yinchuan to develop partnerships with local hotels and restaurants so that visitors would have a place to stay in town. Roads in the town of Yinchuan were improved, along with a beautification scheme, such as adding fountains, trees, and flowers, and removing rundown buildings.

The Ningxia government also began organizing wine tourism events, such as the Ningxia Grape and Wine Festival that was started in the summer of 2012, and has been offered every year since. They also host an International Wine Expo and Site Vinitech that brings in international wine experts, suppliers, and tourists.

They still need to develop a wine tourism website and other marketing materials, such as brochures and maps, to be available in multiple languages. In addition, they will need to hire a professional marketing team to develop a logo and advertising campaign. However, since the strategy is to open Ningxia to international tourist in 2017, there is still time to accomplish this.

RESULTS AND BEST PRACTICE IMPLICATIONS

Ningxia can be considered a best practice in wine tourism because they effectively built a wine region from scratch in a little over a decade. More importantly, they were able to achieve very positive press based on the high quality of their wines. Not only did the 2009 Jaibelan win the International Trophy for best Red Bordeaux Varietal Over £10, many of the wines from other Ningxia wineries have also received glowing reviews. Silver Heights wines, produced by female winemaker, Emma Gao, have received excellent reviews from Jancis Robinson (2012a), and Michel Bettane, Thierry Desseauve and Jeremy Oliver are also impressed with Ningxia wines (Moselle, 2015). Other well perceived Ningxia wine brands include Bacchus, Lanny, Leirenshou, and St. Louis Ding.

The Ningxia wine region grew from a handful of wineries in the mid 2000's to 72 wineries by the end of 2014, with many having very large and impressive architecture. The size of the vineyard land has also increased from 7,800 hectares in 2006 to over 39,000 hectares in 2014, and currently produces wine that is evaluated to be worth 6.5 billion yuan, equating to around one billion US dollars in revenues (Cao, 2015).

The Ningxia winery classification system has resulted in ten wineries earning fifth-growth status in the 2013, as determined by a panel of international experts. These wineries are: Xixia King, Chateaux Yuanshi, Helan Qingxue, Bacchus, Yuange, Changyu Moser XV, Lanyi, Yuhuang, Leirenshou, and Chengcheng (Thach, 2014).

Ningxia has also implemented quality control systems for viticulture and winemaking. Because of this, they are one of only two wine regions in China that have been invited to become an OIV observer. They were also recognized by the Chinese Committee of National Geography with their Geographic Indicator of the East Foothill of Helan Mountains, which was also listed in *The World Atlas of Wine* in 2013.

The region has also illustrated a pattern of cooperation between government, social organizations, such as universities and training centers, and private enterprises. Collaboration between different entities is considered to be one of the hallmarks of best practice wine tourism (Getz, 2001).

In terms of results, the Ningxia region now receives over 1.65 million tourists each year, which spent an estimated 14 billion yuan ($2.2 billion US dollars) in Ningxia in 2014 (Cao, 2015). The Ningxia government estimates it will see a financial return on their investment in developing their wine industry within 7 to 11 years (Cao, 2015).

FUTURE ISSUES

Despite the rapid advancement of the region, the Ningxia wine industry still faces many challenges. One issue is the high level of wine supply outside of China, which has inspired well-known wines from the Old and New World wine producing countries to flow into China at relatively low prices. This puts great pressure on domestic Chinese wine producers.

Another issue has to do with the fact that Ningxia is still a young wine region, and is working hard to implement their strategy to increased quality in vineyards and wine production. Though they have succeeded to some extent, they still need to strive for continuous improvement. They are working to create an integrated industrial chain comprised of world-class vineyard development, wine production, and marketing/sales to ensure long-term sustainability of the wine industry.

Areas on which the Ningxia government knows they still need to focus are:

- **Creating a strong brand for the region.** This includes clarification of the regional logo and emphasis on high quality wine, health, sport, and relaxation as the main messages. There is a need to explore new ways of marketing such as launching an international website, utilizing social media, and establishing retails sales of Ningxia wine in restaurants and wine shops in top cities like Beijing, Shanghai, Guangzhou, and others. In addition, Ningxia needs to continue to submit wines to both domestic and international wine competitions in order to gain feedback and, hopefully, more awards. They also want to develop a Ningxia International Wine Exchange and comprehensive Free Trade Zone.
- **Reinforcing the management of Ningxia Chateaux with Cru Classe Status.** As the strategic orientation of Ningxia wine industry is producing wine of high quality, the items listed in the regulations of the Ningxia Chateaux of Cru Classe system should be strictly followed. This includes stringent international evaluate of the vineyards, wines and chateaux of Cru Classe Status every two years.
- **Facilitation of wine tourism and culture.** Ningxia needs to continue to work on implementing their wine tourism infrastructure with a focus on high quality wine, health, sport, and relaxation. In order to do this, they need to add additional elements of sightseeing, food, shopping, and entertainment into tourism offering. Additionally, they need to expand their base of vineyards, chateaux, restaurants, and sophisticated leisure activities, related to this theme.
- **Continued Development of Skills and New Jobs**. The Ningxia government recognizes it needs to continue to educate workers in viticulture, winemaking, wine tourism, marketing, and construction related to the wine industry.

Finally, Ningxia refuses to rest on its laurels and has expanded its vision for the future. The Comprehensive Planning of the Development of Wine Industry and Cultural Corridor believes that by the year 2020, vineyards in Ningxia will have increased to over 66,000 hectares. In addition to the establishment of a Grape Culture Center, they plan to build three ecological wine cities, ten wine-themed towns of different styles, and reach a level of at least 100 supreme chateaux. Their vision is to become the location of the greatest chateaux cluster in China, the biggest chateau wine-producing area in Asia, and one of the premier wine producing regions of the world.

DISCUSSION QUESTIONS

1. Conduct a SWOT analysis on the Ningxia wine region, identifying Strengths, Weaknesses, Opportunities and Threats.
2. Analyze Ningxia's vision and strategy for the wine industry. How have they managed to achieve so much in such a short time? Will they be able to achieve their vision in the future?
3. What does Ningxia need to do to sustain its success? Identify at least three success factors, and what they need to do to implement them.
4. Discuss the differences between wine tourism in China and other regions of the world, such as Bordeaux. What can the different regions learn from one another?
5. Assume you are in charge of wine tourism for Ningxia. What steps would you take to ensure a steady increase of tourists and revenue to the region in the next ten years, and achieve positive sustainability (respect for environment, equity of people, and economic return)?

Chapter Five

"Toujours Bordeaux!"
The Creation of a Cultural Wine Center

Julien Cusin & Juliette Passebois-Ducrosj
IAE School of Management, Bordeaux, France

The marketing director of a regional tourism agency in Aquitaine summed up the situation perfectly. "In a city like Bordeaux, the fact that there is no place that is dedicated to wine is strange for everyone! When people arrive in Bordeaux, they are often surprised that there is nothing about wine. It's a very strong image, known all over the world. When you tell people abroad that you come from Bordeaux, they immediately say, "Ah Bordeaux... the wine!" So when tourists come here, they say, "Hey, I'm in Bordeaux, what can I see about wine?" The idea that such a reputation must include a tourist venue to visit seems obvious."

It may seem apparent to many wine enthusiasts around the world that the name "Bordeaux" immediately conjures up images of wine, but to the city planners, until recently, this was not the case. While Bordeaux is one of the oldest and most famous winemaking regions in the world, the Bordeaux area is not at present, an international wine tourism 'hub'.

However, in the past several years, the concept of building a Cultural Wine Center (CWC) emerged, and was eventually approved. The venture is the outcome of the determination of the mayor, Alain Juppé, to turn the city into a thriving regional capital. The issues at stake are crucial for Bordeaux in view of the fact that the wine sector has been hard hit by the financial crisis and has been forced to apply for public assistance.

Thus, it has become urgent to promote wine tourism in Bordeaux and the surrounding area. Moreover, the present context appears to have made such a turnaround possible. In the last twenty years, the town has undergone a major facelift and is now ready to welcome a cultural flagship.

To this end, the "Cite des Civilisation du Vin" is scheduled to open to the public in Bordeaux in 2016. An examination of the factors leading to this wine tourism solution, including background on the region and how the problem presented itself, is described in this chapter.

OVERVIEW OF WINE IN FRANCE

Introduced in France in the 1st century before Christ, wine is more than a simple beverage; it is a cultural product, and a major aspect of the French heritage. This long tradition has given French wines an enviable reputation worldwide. Domestically, this ancient winemaking tradition has resulted in the widespread consumption of wine: 85% of French people buy wine for their personal consumption (France Agrimer, 2010), French consumers are some of the largest consumers of wine in the world with about 44 litres per inhabitant per year (Bonial, 2014), and France consumes around 14% of global wine production (OIV, 2012). In France, wine is both a popular product that is sold and consumed daily, as well as a luxury beverage that is shared with friends and family on special occasions. Wine tasting is something that is passed down from one generation to the next, and involves a specific learning process.

Moreover, wine is a major economic sector in France with a turnover of over 11 billion euros in 2013, representing the second surplus item in the French balance of payment up to 2013 (FEVS, 2013). The wine industry not only creates economic value but also supports around 550,000 jobs, both directly and indirectly (Vin et Société, 2013). According to the OIV (2012), France is the leading wine producer in the world at around 16% of global production, with 42,000,000 hectolitres of wine produced in 2013, from 804,687 hectares of vineyards in France.

There are sixteen wine-producing regions in France, each area giving rise to specific wine aromas and flavors. This wide variety of wines is also linked to the enormous number of small producers and châteaux or estates, with over 85,000 wine producers (FranceAgriMer, 2013).

Wine Tourism: France Catching up with "New World Countries"

While France is a major wine producer and a leading tourist destination, with 81 million tourists in 2013 (DGE, 2014), wine tourism is currently underfunded, poorly structured, and consequently underdeveloped. Conversely, many of the "new world" wine producers – Australia, California, and South Africa – have incorporated wine tourism into their wine offer, achieving excellent economic performance (Lignon-Darmaillac, 2009).

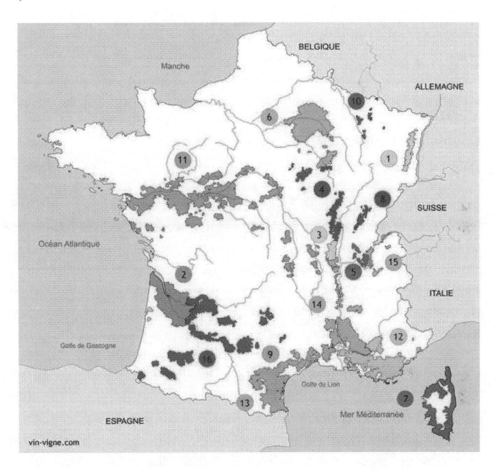

1-Alsace	5-Bugey	9-Languedoc	13-Roussillon
2-Bordeaux	6-Champagne	10-Lorraine	14-Rhône
3-Beaujolais	7-Corse	11-Loire	15-Savoie
4-Bourgogne	8-Jura	12-Provence	16-Sud-Ouest

Map of Major French Wine Regions

The development of wine tourism is relatively recent in France, with the first wine trail created in the Alsace region in 1953 (Lignon-Darmaillac, 2009). Furthermore, though France receives many wine tourists, it is rarely ranked among the top wine tourism destinations in the world (Rimaud, 2011). Successive French governments have been working to develop this potential opportunity, and wine tourism is now recognized as a strategic issue for the agricultural (César, 2002) and tourist sectors. In 2009, a wine tourism committee was set up by French ministries, Conseil Supérieur de l'Oenotourisme, in order to *"implement a new dynamic for wine tourism in France."* (French ministry of Economy, 2009). Consequently, policies and plans have been rolled out to encourage wine professionals to turn to wine tourism, to extend the wine tourism offer and, globally, to enhance France's image as a wine tourism destination.

Overview of the Bordeaux Wine Region

The Bordeaux wine region is more than 2000 years old, and is the largest in the world with 117,000 hectares under vine (Vin-Vigne, 2015). It specializes in making both red and white Bordeaux wines, with the red wines primarily using cabernet sauvignon, merlot, cabernet franc, and petite verdot grapes, and the white wines focusing on sauvignon blanc and semillon grapes. In addition to dry red and white wines, the region of Bordeaux also produces rose, sparkling, and the famous dessert wines from the Sauternes and Barsac regions.

Bordeaux is divided into 54 appellations and boasts 7375 wineries (Leve, 2015). Traditionally, Bordeaux has been divided into six major regions, each with their own specific features:

1. **The Medoc** region (Medoc, Haut Médoc, Saint Esteph, Pauillac, Saint-Julien, Margaux) is undoubtedly the most prestigious since four of the five first growth wineries from the official classification of Bordeaux wine in 1855 are located here. The area is popular with tourists, and about 20% of trips made by wine tourists are in this area (CRT Aquitaine, 2013). Pauillac, the main village of this region, is about 52 kilometers north of Bordeaux city. Due to its position along the river, it is known as the Left Bank, and produces blended wines that are dominated by cabernet sauvignon.
2. **The Graves** region (Pessac-Leognan, Graves, and Graves Supérieures) is especially known for its "dry white Bordeaux" wines, with a distinct mineral-y taste and refreshing acidity. However, it is also home to the famous Chateau Haut Brion, which makes both red

and white Bordeaux, and is one of the distinguished five first growth wineries. A Graves wine trail has been developed, and the area is often visited for its gastronomic restaurants.

3. **Entre-deux-mers**, which means "between two seas," is a large area with 1500 hectares located between two rivers, the Garonne and the Dordogne, hence its name. The area is visited both for its refreshing and fruity white wines and its heritage sites that include a large number of abbeys, Romanesque churches, and fortified bastides.

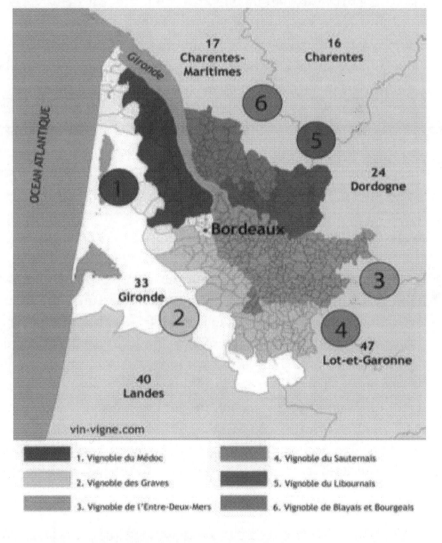

Map of Bordeaux Wine Regions

4. **The Sauternais** region (Sauternes and Barsac) is famous for its sweet dessert wines made from sauvignon blanc and semillon grapes. A Sauternes wine trail has been developed, with tourists enjoying the hilly countryside and beautiful old chateaux of the region. The tiny village of Sauternes is located 48 km south of the city of Bordeaux.

5. **The Libournais** is a famous region linked to magic names such as "Petrus", "Cheval Blanc" or "Château Ausone". The area is also known for its heritage sites such as Saint-Emilion, a lovely village listed as a UNESCO world heritage site. This nice village is an approximately 40-minute drive from Bordeaux city. Since it is across the river, the region is also known as part of the Right Bank, which focuses more on the merlot and cabernet franc grapes.

6. **The Blayais and Bourgeais** region is a hilly area that boasts ideal exposure for vines, guaranteeing a **healthy and fully ripe crop**. This region is less well known, although it also enjoys a good reputation and is visited for its citadel constructed by Vauban and listed as a UNESCO world heritage site. The main city of this region, Blaye, is about 50 kilometers from Bordeaux city. This is also known as part of the Right Bank.

Wine Tourism in Bordeaux

In spite of its long wine-producing history, the Bordeaux region has only recently taken up the opportunity presented by wine tourism. According to Lignon-Darmaillac (2009), this is firstly due to the dominant "wine merchant oriented culture" in Bordeaux that cut Bordeaux wine producers off from their customers. Moreover, Bordeaux wines have largely been developed with respect to tradition with a focus on excellence that doesn't always take consumer needs into account. Consequently, even though the Bordeaux region is a very popular tourist destination, with Aquitaine listed as the 5th largest tourist traffic region in France, only 10% of foreign visitors come to Bordeaux specifically for its wines (CRT Aquitaine, 2013).

# of Wineries in France	85,000
# of Wineries in Bordeaux	7,375

Many tourists opt to stay on the coast in the Arcachon Bay, or visit Bordeaux's cultural heritage or the caves in Périgord. According to the CRT

Aquitaine (2013), the region hosts almost 2.7 million wine tourists during the high season, 39% coming from abroad, mainly from European countries, with 4% from the USA. Those tourists who come to Bordeaux for wine are usually highly involved wine consumers who have visited at least 3.3 other wine regions in the last five years. They generally have high incomes, and spend much money on wine, gastronomy, and tasting.

THE PROBLEM: NO CENTRAL LOCATION FOR WINE TOURISM

The French city of Bordeaux is in a highly paradoxical situation, because, on one hand, Bordeaux has an international reputation and real prestige for the high quality of its "Grand Crus" wines. Its most legendary wines, such as Château Pétrus, Château Mouton Rothschild, Château Margaux, Château Latour, Château Lafite Rothschild, Château Haut-Brion, Château Cheval Blanc, Château d'Yquem, and Château Ausone serve as references for all the great wines in the world (La revue du vin de France, 2012).

Iconic Wine Label from the Bordeaux Region

Furthermore, many of the widespread wine-related practices are linked to the history and tradition of the wines of Bordeaux: the Bordeaux color, the concept of a wine château, the unit of measurement of the barrel (900 litres), the notion of vintage, and Bordeaux's linkage to the origin of the creation of the AOC with the concept of wine classification (Inno'vin, 2006). In addition, Bordeaux organizes a famous biannual trade show called *Vinexpo* for professionals from all over the world, with an average attendance of around 50,000 people.

On the other hand, Bordeaux's wine tourism offer lacks organization and structure. Consequently, the situation is very complex for tourists that wish to discover the world of Bordeaux wines. Indeed, Bordeaux is a large city with about 240,000 inhabitants, but has no wineries or properties producing wine downtown. The iconic names – Margaux, Saint-Emilion, and Sauternes – and the beautiful properties and châteaux that have given Bordeaux its reputation are located between 30 and 50 kilometers from the city.

Moreover, not every region benefits from a great reputation, with the Medoc and Libournais together generating more than half of all wine visits (CRT Aquitaine, 2013). Furthermore, there is no visible office or emblematic venue that guides tourists from the city to the vineyards and châteaux, and no standard tourist route. Generally speaking, there is a perceived lack of information about wine tourism and this increases the elitist reputation of Bordeaux wines. As a result, the city widely considered as an international reference in terms of wine, actually does not do a good job at supporting the average wine tourist.

Following are some quotes that emphasize the problem:

> *"I met an elected politician and I told him: "I'm surprised that you have nothing on wine worthy of the name'! When an American or a Japanese person is in Bordeaux, they expect two things. First, they expect a cultural center based on wine and then they expect a coordinated visit of the châteaux. We need to be positioned once and for all as "THE" world capital of wine. Bordeaux is international! It's worldwide. People don't even know where Bordeaux is on a map, but it's a myth and it's extraordinary!"* (Chief of staff to a politician in Bordeaux, 2012)

> *This may seem surprising, but the world's wine capital has never had a museum worthy of its reputation"* (Le Monde, 01/08/2011).

> *"We've been seeing the emergence of tourism in Bordeaux in the last couple of years. The city has a strong brand. It has an international reputation that's based on wine. Wine is something that attracts tourists*

and there are many visitors who come for it. In these circumstances, the fact that there's no place to talk about wine – i.e. a place for explaining and showing Bordeaux wine – it's a kind of aberration!

Today, it's undeniable that wine and tourism are very important assets for the city of Bordeaux. When foreign visitors come to Bordeaux, they discover a heritage that they didn't expect. But very soon their question is: "And the wine?" They come to the Tourist Information Center, asking us: "Do you have a sort of cathedral, a museum, or a Guggenheim of wine?" And we tell them to go to the vineyards of Bordeaux.... Bordeaux is part of the history of wine!" (President of the tourist information center, 2012)

THE SOLUTION: CREATION OF A CENTER FOR WINE TOURISM IN BORDEAUX

In 2008, the local authorities finally agreed to launch a huge scheme devoted to wine culture in the city. They had tried to launch a similar project in 1995, but it was abandoned for numerous political and economic reasons (Cusin and Passebois, 2014). However, the year 2008 provided a favorable context: Bordeaux was undergoing a major facelift that lasted for 15 years with the redevelopment of wharves, refurbished façades, and reduction of bottlenecks by the introduction of a tramway. The improvements have been so well received, that Bordeaux is now considered to be one of the best cities in which to live in France (OpinionWay, 2014).

Furthermore, the number of tourists had also increased when Bordeaux became a UNESCO world heritage site in 2007. Finally, the mayor, Alain Juppé, had decided to position the city as a wine capital of the world, and was drawing up an application to compete for the title of "European capital of culture 2013". Therefore, the timing was right, and the city was ready and committed to welcoming a center dedicated to wine culture: the "city of wine civilizations" (CWC).

A Three-Stage Process

Expected to open in 2016, the project has gone through three major stages:

1) ***Commitment of the Mayor*** - the first stage involved the personal commitment of the mayor, who made it a priority for the city.

The mayor had to convince other local authorities from opposing political camps to finance the project. Step by step, driven by determination, he was able to "coerce a consensus" on the project.

2) ***Support of the Wine Industry*** - The second step was to draw wine professionals into the scheme and to encourage them to embrace it. To this end, the mayor called Sylvie Cazes to join the Bordeaux town council in charge of promoting wine and wine tourism sectors. Sylvie is a wine producer, and formerly President of the Union of Grands Crus of Bordeaux (a highly influential 'club' which includes 130 of the biggest Bordeaux vineyards). Closely linked to the world of wine, her presence meant that the project was able to get the sector's backing.

3) ***Formation of a Project Team*** - Third, a nine-person project team was formed, including a cultural director, science manager, and development manager. It was headed by a former operational director of a major French tourist site, Philippe Massol.

The collaboration of experts in cultural matters, tourism professionals, wine makers and producers was a key factor in making the project a success. The structure brought together all the institutional players who would subsequently be involved in the project, including the Bordeaux town council, the Communauté Urbaine de Bordeaux (CUB), the Regional Council of Aquitaine, the Committee du International Vin du Bordeaux (CIVB), and the Bordeaux Chamber of Commerce and Industry (CCI)

The Director of the Bordeaux Wines Inter-Profession explained why the project was successful:

"The project was put back on the agenda at a time when Bordeaux, prompted by Alain Juppé, wanted to compete for the title of "Bordeaux, European capital of culture". The mayor of Bordeaux wanted us to forge ahead with the center, even though the other local authorities were somewhat lukewarm. It's clear that you need a political leader who can drive the project through!

From the moment it becomes part of a larger project, in other words, the "Bordeaux, European capital of culture" project, and you shift from being a small scheme of a few million euros to something which amounts to more than 60 million euros, and, behind it all, you have a true driver of the project who says "whatever happens, I'll see it through to the end," at that point your mind set changes completely. And then you give yourself the resources to reach the goals you've set."

The Architectural Elements of the CWC

Gradually, an ambitious project, worth about $81 million euros in investment, emerged from the team. It included a large building of 15,000 meters with an audacious architectural design that focused on the wine civilization in general, and not simply on Bordeaux wines. The design has the air of a futuristic setting built in a fully restructured neighborhood. It became known as the Cultural Wine Center (CWC) and was called the "Cite des Civilisation du Vin."

Architectural Design of Cité des Civilisations du Vin

Anticipated attendance is around 425,000 visitors per year and the CWC is expected to make $50m in revenue and to create 750 stable jobs. According to the website of the CWC (2015) it is *"neither a museum nor a theme park, but an original concept which lies somewhere in between."* The project benefits from the backing of the wine sector professionals, and virtually all the local authorities.

In designing the CWC, the local authorities wanted to create a "must see" for tourists visiting Bordeaux city and its region. The CWC is going to be the "emblem" of Bordeaux city, just as the Eiffel Tower is in Paris, the

Guggenheim Museum in Bilbao or Big Ben in London. It will be featured in advertising campaigns, tourist guides and on postcards.

Alain Juppé, mayor of Bordeaux, explained that Bordeaux is constructing a "world-class facility that embodies its title of world's wine capital."

RESULTS AND BEST PRACTICE IMPLICATIONS

As the CWC is not expected to open until 2016, it is difficult to claim that it is a resounding success already and a model to follow. However, an analysis of its emergence and the project description suggest that the conditions are in place for it to be a success, not only in its stated goals and ability to meet its attendance rate, but also in the enhancement of wine tourism in the Bordeaux area. Indeed, the CWC is the outcome of a regional strategy designed to promote wine tourism.

The project addresses the new challenges of venue marketing and may be considered as "best practice". There are four key success factors in this regional wine-tourism promotion strategy:

Factor One: Creation of a Cultural Flagship

First of all, the CWC is a masterpiece in city branding, designed to consolidate the significant awareness and reputation of Bordeaux wines by opening a structure dedicated to wine culture, tradition, and tasting in the center of the city. The CWC is a unique concept, but it belongs to the so-called "cultural flagship projects": "*high-profile, multi-use, and often large-scale arts facilities typically designed by world-renowned architects and endorsed as among a city's most spectacular attractions*" (Grodach, 2008:495). Its high cost, attendance rate, size, and disruptive design, make the CWC a decidedly ambitious endeavor and a cultural flagship.

"Numerous studies have highlighted the consistence of regional strategies, modelled on the Guggenheim Museum Bilbao, that use cultural flagships to improve their image and their attendance (Plaza, 2009; Gonzales, 2011). Cultural flagships can be successful if, on one hand, they manage to combine a cultural and a consumer experience and, on the other, they are easily visible and memorable. An original architectural design is a key success factor in making the site visible to tourists and enhancing public awareness. The CWC is not a museum to "learn about wine culture" but is a place where visitors can experience wine, taste it, smell it, have fun, and dream (Schmitt, 1999). The CWC will offer visitors a unique, memorable, and multi-sensory experience.

Factor Two: City Branding as World Wine Capital

Second, Bordeaux is involved in a branding strategy, which means it wants to turn the city into an attractive tourist destination, while also making it attractive to investors, start-ups, and inhabitants. To brand a city, decision-makers need to *"introduce a certain order or coherence to the multiform reality around us, enable us more easily to 'read' each other and our environment of places and products. Brands are not purely a source of differentiation, but also of identification, recognition, continuity and collectivity"* (Mommaas, 2002:34).

The CWC has been designed to serve Bordeaux's branding and to turn the city into the "world wine capital". According to the CWC website (2015), both in its content and its shape, the center has been inspired by wine: *"the building is a pure evocation of the spirit of wine and a unique place where city and river meet"*. It is entirely focused on the wine experience. As its designers explained, the center will give visitors an enjoyable experience, based on interactive tools, and will offer a *"voyage of discovery through the history of wine and civilization, exploring various facets of this ancient relationship."*

The purpose is not to speak about "Bordeaux wines" but to encompass the wine civilization. Thus, visitors will discover the wine civilization from different perspectives through 17 "wine universes", including the historic dimension, the geographical aspect, the sensory dimension, the technical aspect of wine and the economic side. Therefore, the value proposition developed in the center is about the culture and civilization of wine in general and is not limited to the traditions of Bordeaux wines.

Factor Three: Serving as a Tourist Information Center

Thirdly, in addition to the cultural experience, the CWC will be an information center for tourists. There will be numerous documents and guides within the building to inform people about the diversity of vineyards, châteaux, and properties surrounding the city. The CWC has been designed as a "platform for guidance" to enable tourists to identify other available wine-related resources and so enhance their experience.

The model proposed is a mix between a "must see" downtown activity and a "tourist platform" that will provide tourists with a plethora of information on wine tourism activities. The CWC will be a place of convergence for different agencies promoting wine, tourism, and wine tourism.

The idea of gathering actors under one roof that work separately from one another is, in itself, a form of best practice. Moreover, the facility responds perfectly to the challenges Bordeaux is faced with. On one hand it is a well-known city, famous for its wines but without the facilities to experience wine and, on the other, it has a huge vineyard, far from the town center, that can deliver rich wine tourism offers in the area where the wine is produced. The CWC has addressed this challenge and will act as a key driving force to develop wine tourism over a large area.

Factor Four: Collaborative Partnerships & Leadership

Finally, the CWC is the outcome of a 20-year project that involved a large number of local actors and authorities. From 1995, there were three major setbacks before the project of a wine-based cultural and tourist facility finally took off. The emergent project will not only help to boost sales and exports of Bordeaux wine but will also boost the entire wine industry. It is a collaborative partnership that involves both wine sector actors (producers and trade unions) as well as promoters of local tourism.

However, it is important to keep in mind that the project would never have been possible without the perseverance and leadership of the mayor of Bordeaux. Indeed, this success story is a good illustration of the approach that a decision-maker must adopt if a project he or she supports fails. Indeed, it would appear that persisting with a project is pertinent when the choice is driven, from the outset, by a clear strategic vision on the part of the decision makers (Lynn & al., 1996).

In the present case, the mayor of Bordeaux immediately saw the potential of a CWC for the Aquitaine wine industry. When the project ran up against difficulties, he decided to publicly show support for the project and to personally engage in the project's success. Despite the former failures, his political determination removed some of the uncertainty surrounding the project, and consequently legitimized the increased commitment of all the stakeholders. It was only when the mayor of Bordeaux upped the stakes that the CWC achieved the success that had previously eluded it.

In short, if a project is judged as interesting but is initially a failure, increased commitment should not necessarily be viewed as a form of escalation (Staw, 1976), but may instead be seen as a strong political sign of the project's importance in the eyes of the decision-makers, subsequently eliciting greater enthusiasm from the other stakeholders.

FUTURE ISSUES

The CWC may well be successful, but it cannot be considered as the only lever to consolidate Bordeaux's role as the world capital of wine. Indeed, many challenges remain for the Bordeaux region, including those that were identified by the regional wine industry cluster (Inno'vin, 2006):

Understanding Expectations of New Wine Consumers: The Bordeaux wine industry must meet the expectations of a growing number of consumers, with new consumption habits, new global consumers, demand for high quality, etc. In other words, the Bordeaux region has to reconcile tradition and innovation to attract new customers with new expectations.

The opening of an innovative venue with an ultra-modern design can, as such, change the image of Bordeaux and its vineyards. Attracting consumers from around the world in one single place is also an effective way to understand their expectations. In the future, it will also be necessary to bring together the wine industry sector and the tourism sector, which have worked separately in Bordeaux for too many years.

Competing in the Global Market: The wine industry in Bordeaux faces increasing global competition since the entry on the market of *new wine-producing countries such as* Australia, Chile, the United States, and others, which generally have a powerful marketing orientation. Consequently, Bordeaux needs to promote its wine at international level, telling the story of a high quality and traditional brand. The CWC is the main tool for this promotional strategy. Consequently, the center will serve to enhance the exportation of Bordeaux wines thanks to the links created with foreign tourists during site visits.

Increased Collaboration Amongst Bordeaux Wine Sector: The Bordeaux wine industry is a highly fragmented sector and needs to evolve so that the various initiatives in terms of production, distribution, and marketing are less diluted in the future. It is especially important to structure the wine tourism offer collectively, such as setting up tours in the Bordeaux vineyard.

The creation of the CWC as a symbolic venue for all existing vineyards is a first step in this direction, and the *vineyard indicator platform* concept has been specifically designed to make the center a gateway to the Bordeaux vineyards. The recent creation of a wine tourism application for smartphones and tablets created for visitors to the Bordeaux vineyards fits into the same logic.

Adopting Sustainable Production Methods: - The wine industry in Bordeaux operates in a field in which the environment has become a marketing argument with the emphasis on sustainable development. Thus, the CWC project adopts a sustainable development approach, not least because it will open in a new eco-district in Bordeaux, where the latest energy-saving systems will be tested.

In short, the CWC should be considered as a necessary step for the Bordeaux region to adapt to the different challenges it faces. There was clearly a need for creativity in the Bordeaux wine industry that the CWC has undoubtedly catalysed.

DISCUSSION QUESTIONS

1. What advice would you give to local politicians seeking to develop wine tourism?
2. Once local politicians have chosen the solution of the "*cultural flagship*", what existing international facilities could inspire them?
3. What content can we visualize in the creation of a cultural and tourist center about wine?

Chapter Six

Is Good Wine Enough? Place, Reputation, and Wine Tourism in Burgundy

Laurence Cogan, Steve Charters, Joanna Fountain,
Claude Chapuis, & Benoît Lecat
Burgundy School of Business & Lincoln University

Marquise Caroline de Roussy de Sales was delighted when she welcomed us on a foggy day last November in her château. She had just received a national prize for the chateau's gardens, designed by famous landscaper Lenôtre in 1676. Today, visitors come from all over the world to admire the beautiful rose-garden adorned with 220 bushes of 15 varieties. Then, they taste the wines of the estate and visit the wine cellars, which are the longest ones in Beaujolais, the southern more region of Burgundy, France. Château de la Chaize is the largest domaine in Beaujolais with 99 hectares in the Brouilly appellation. However, when asked about wine tourism, the Marquise gave an ironic answer:

> *"Wine tourism is a double-edged sword; whether you receive*
> *a couple or a group of forty people you will end up spending the*
> *same amount of time and energy. Wine tourism is time-consuming,*
> *but it is a way to get our wines known".*

In contrast, Julie Leflaive, daughter of Olivier Leflaive strongly believes in wine tourism. Her father, who opened the first "table d'hôte" in Puligny-Montrachet about twenty years ago, was a pioneer in wine tourism in Burgundy. Now, they offer a complete experience in wine tourism, including a tour of their vineyards in French and in English, a visit to their cellars, a wine tasting lunch or dinner, and a boutique hotel. Olivier Leflaive, unlike many wine-growers in Burgundy, was always a strong supporter of wine tourism. According to his daughter, one of the reasons he supports wine tourism is because "most of his customers are English-speaking and become his best brand ambassadors when they return home to their country". He works hard to build the capacity of the district to welcome wine tourists, focusing not just on his business but on collaboration with others.

These key players in the world of wine tourism in Burgundy introduce us to the paradox of wine tourism in France, especially in these two sub-regions. From a historical point of view, as both regions belong viticulturally to Burgundy, Beaujolais, and the Côte d'Or have enjoyed a high reputation for the last 100 years, thanks to developing road transport throughout the region. Yet despite the renown of the wines, wine tourism in both sub-regions seems to be under-developed, though they have a huge potential to make the most of it.

This chapter is unique in that it doesn't present Burgundy as a best practice in global wine tourism, but does recognize that Burgundy is world famous to wine tourists because of its reputation for fine wine. Indeed, the fact that eight of the top ten most expensive wines in the world come from Burgundy (Young, 2013) supports the premise that Burgundian wine is a luxury product (Beverland, 2006), and has achieved a brand position to which many other wine regions of the world aspire.

The questions posed are, "Is good wine enough?" and "Does Burgundy need to enhance its wine tourism offering?" By examining the strengths and weaknesses of wine tourism in the Cote d'Or and Beaujolais regions of Burgundy, this chapter identifies problems and provides a series of recommended solutions on additional steps Burgundy should take in order to move into the realm of a true best practice destination for wine tourism. Based on a series of in-depth interviews with 23 winemakers, négociants and tourism authorities from Burgundy, this chapter provides deep insight into wine tourism issues that can arise in a region that has a long heritage based on luxury wine.

Wine Tourism in France - the French Paradox

According to the French Ministry of Tourism (2013), France is the world's top tourist destination, with about 83 million foreign tourists in 2012. Its long history of wine-production goes back to the Romans and in

2013 it was the second largest wine producing nation, with 44 million hectoliters, behind Italy, but well before Spain and the USA. It undoubtedly has the most renowned wines in the world, and the reputation of being a mecca for wine lovers.

In spite of these factors, wine tourism in France is clearly under-developed. The "Dubrule Report" published in 2007 and commissioned by the French government exposed a clear under-development of wine tourism in French wine regions. This report identified some key issues:

- A lack of cooperation between the hospitality industry and the wine industry;
- A lack of knowledge about the wine tourist profile;
- A lack of professionalism in welcoming wine tourists;
- An uncoordinated private sector, with some successful sporadic private wine tourism initiatives.

These general concerns are reflected in Burgundy. Research suggests that, particularly for lower-involvement non-French tourists, the region assumes that the quality of wine is sufficient to entice and satisfy the visitor (Charters, 2015). The same research reveals that overall service, however, is perceived to have major flaws.

OVERVIEW OF THE BURGUNDY WINE REGION

Burgundy is a French administrative region comprising four *departements*, or counties, in central eastern France. Located about two hours south of Paris, the region of Burgundy is comprised of six major wine sub-regions. Starting from the north to the south, the sub-regions are:

1) *Chablis* – known for its steely chardonnays and cooler climate
2) *Cote d' Nuits* – most famous for its velvety pinot noirs and some of the most expensive vineyards in the world, including La Tache and Romanee Conti. The Cote d'Nuits is the northern part of the Cote d'Or, which means "Golden Slope."
3) *Cote d' Beaune* – most famous for its voluptuous and long-lasting chardonnays, and the well-known vineyards of Le Montrachet and Le Batard Montrachet. The Cote d'Beaune is the southern part of the Cote d'Or.

4) ***Cote Chalonnaise*** – known for its leaner elegant wines and Crémant sparkling wines
5) ***Mâconnaise (Mâcon)*** – most famous for its broader buttery chardonnays and the village of Pouilly-Fuissé
6) ***Beaujolais*** – known for its fruity gamay wines and beautiful old chateaux. Though not in the administrative region of Burgundy, it has always been regarded viticulturally as part of Burgundy. Indeed Beaujolais and the rest of Burgundy are united by a common appellation, AC Côteaux Bourguignons which traverses the two sub-regions.

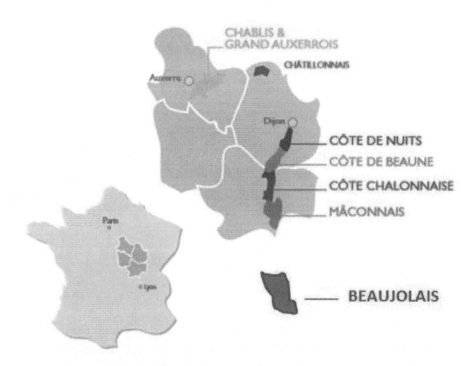

Map of Burgundy Wine Region

History of Burgundy Wines

In Burgundy, wine cannot be separated from history. The oldest known vineyard to-date, recently discovered in Gevrey-Chambertin, was planted in around 50 A.D. For 600 years, the monks of Cîteaux owned the château of Clos de Vougeot and the role of the church in developing the wine industry and promoting its wines was fundamental. The wines of the Côte d'Or were already famous in the Middle Ages and maintained their reputation over the centuries which followed. After the French revolution, vineyards belonging to aristocrats, and religious orders were sold to a few rich merchants and a multitude of small growers. Merchants, known as *négociants*, became important. They bought wine in bulk from small growers and sold it on the home market and abroad – often blending it. To some extent, this system still exists today.

Hospices de Beaune in Burgundy

Throughout history, Burgundy faced serious crises. During the depression of the 1930's, Burgundy producers discovered the importance of public relations. During this time the brotherhood of the Chevaliers du Tastevin was created - a *confrerie* of those who love the wines of the regions and celebrates them with substantial dinners and major events. This has attracted artists, writers, scientists, and politicians to Clos de Vougeot and incited them to drink Burgundy. These famous guests in turn became ardent propagandists for the wines. Likewise, for the past thirty years, the century-old wine auction held in the Hospices de Beaune has become a major event covered by the media worldwide. It is a unique and formidable shop window for the province.

Today, many properties still remain in the hands of families. They are small, averaging 4 to 6 hectares. In the 1970s, a new factor appeared: direct sales from the estate. An increasingly larger proportion, around 15%, of the wine is sold this way, and now forms the backbone of wine tourism in the region (BIVB, 2014).

For over a millennium, most vineyard work and winemaking in Beaujolais was carried out by Benedictine monks. The region was named after Beaujeu, and was overseen by the Lords of that town until it was annexed by the Dukes of Burgundy in the 15th century. The regions' wines were mainly sold in markets along the Rhône and Saône and especially in the city of Lyon. The growth of French railways early in the 1800s gave producers access to the rich market of Paris.

Subsequently the wines won their first reference in English, with the author Cyrus Redding noting that the wines of Saint-Amour and Moulin-à-Vent in Beaujolais were lower priced and needed to be drunk while youthful. The Beaujolais wines were never perceived to be of the same quality as those of the Côte d'Or, but were popular and easy to drink. In the 1950s a new style of wine called Beaujolais Nouveau was developed from the gamay grape. It is young, fruity, inexpensive, and released each year on the third Thursday of November. It gave Beaujolais global attention, and to this day is still the reason for many restaurants, bars, and individuals to have a Nouveau Beaujolais party and buy cases of the wine. Overall this gave the region of Beaujolais an extra impetus, though in the long term may have started to undermine the region's reputation for quality (Garrier, 2003).

Since 1930, Beaujolais vineyards have been deemed part of Burgundy based on a court decision. In 1937, the regional Burgundy Appellation Contrôlée was created including red gamay from Beaujolais. From a wine business point of view, the Beaujolais wine region is linked to the Côte-d'Or due to the business activities of the Beaune négociants, who all bottle and sell wines from the area. However, the Wine Industry Professional Associations of Burgundy and Beaujolais are still independent.

Burgundy Grapes & Appellations

The wines in Burgundy are made with without blending different grape varietals together, as is common in Bordeaux, the Rhone and other French winemaking regions. The dominant grape varieties in the Côte d'Or are Pinot Noir for red wines and Chardonnay for white wines. There is also some Gamay for the reds and some Aligoté for the whites.

According to the BIVB (2011), there are approximately 3800 wine growers in Burgundy. There are 100 appellations, or *protected designation of origin (PDOs)*, in total in the Burgundy region, but it becomes more complicated as the appellation system is based on a 4-tier hierarchical ranking. From top to bottom this includes *grands crus* (1.4%); Premiers Crus, (10%); Village, or communal, appellations, (36.6%), and regional appellations, which represent 52% of the production. Burgundy, especially the Côte d'Or, has unique characteristics: each plot is called a "lieu-dit" and it represents "a piece of land whose name conjures up a topographic or historic specificity" (Pitiot and Servant, 2010 p. 87).

# of Wineries in France	85,000
# of Wineries in Burgundy	3,800

Beaujolais wines are made overwhelmingly from Gamay, a grape also used in the Côte d'Or, although it has secondary importance there. Located along the Beaujolais hills and overlooking the valley of the river Saône, the wine region has developed a tradition of wine exports. Whereas the Côte d'Or is characterized by myriad appellations, Beaujolais produces large quantities of a fairly standardized product. Along with Beaujolais Nouveau, it is possible to find standard Beaujolais wine, including Beaujolais Supérieur, Beaujolais Villages, and the region's highest-quality wines, the ten Beaujolais "crus". Each of these crus has its own appellation title but unlike the crus of the Côte d'Or, they encompass an entire village, not just individual vineyards. They are the best wines produced in Beaujolais and can age on average between 3 and 7 years or even more (Frangin, 1994; Lutun, 2001).

In terms of volume, Beaujolais Nouveau is the most important appellation and accounts for about 5 million cases (Inter Beaujolais, 2014). For many years, the "Beaujolais family" has lived in the shadow of this wine, which was a master stroke of marketing. However, consumer tastes have evolved. Trends go out of fashion and Beaujolais Nouveau is not often seen to represent high quality - though there are exceptions to this, such as the market in Japan (Garrier, 2003).

One of the key features of the culture of wine production in this part of France is the use of the idea of terroir. Terroir currently refers to the typicity of the wine, describing the degree to which a wine reflects its origin and thus demonstrates the signature characteristics of the area where it was produced, its mode of production, and its grape variety (Demossier, 2010). The quest for typicity in an appellation wine was highlighted by its taste related to terroir, a guarantor of the originality of the wine. It is not by chance that the notion of terroir was then highlighted in regions where wine estates have always been small and where wine-growers often make their own wine. Terroir can be used to develop wine tourism activities, but can be more exploited especially given the links between history, culture, craftsmanship, food, and wine.

Famous Vineyard of Romanee Conti

Gastronomy is also a significant aspect of the culture of this part of France. The last century has seen an emphasis on the character of regional cuisine, with specific local ingredients and regional dishes, such as cheeses, mustard, charolais beef, poulet *gaston gérard*, kir and many others. Food and gastronomy has thus become another key attraction of the region.

Additionally, there are many historical tourist attractions, such as the world heritage listed monasteries, old towns, and museums. These also attract a number of visitors to the region.

Strengths of Wine Tourism in the Côte d'Or and Beaujolais

As well as the points noted above about wine reputation, history, and gastronomy, there are some clearly identifiable strengths for wine tourism in both the Cote d'Or and Beaujolais. There is a vast choice of unique tour option, such as cycling tours, visits in carriages, jeeps, hot air balloons, and mopeds, as well as hiking trails and cultural activities. Both regions have a rich architecture including many chateaux. Additionally, there is the cathedral and palace of the Dukes of Burgundy and the Hospice de Beaune in the Côte d'Or as well as, the Golden Stones area and Romanesque Churches in Beaujolais. The "Route des Grands Crus de Bourgogne", established in 1934, is well-signposted in the Côte d'Or and is visited by 3.5 million visitors a year. Meanwhile, Beaujolais benefits from varied landscapes including hills, forests, and beautiful panoramas around the crus villages. The village of Oingt in Beaujolais is also considered to be one of the most beautiful villages in France.

According to the BIVB (2014), wine tourism in Burgundy has improved in the past several years, with 93% of wineries officially open to the public. In general wine tasting is free, but wine tourists are expected to buy some bottles at the end of the tasting. During 2011, wineries sold 30 million bottles direct to consumer at cellar doors in the administrative region of Burgundy, which represents 15% of the region's the total production. This has increased by 3% compared to 2010. Twenty one percent of the wineries hired additional staff to handle visitors in 2011 compared to only 18% in 2010. More wineries now offer vineyard visits, increasing to 63% in 2011 against 55% in 2010. The BIVB (2014) also reports that accommodation is now offered by 13% of wineries, an increase from only 7% in 2010. Furthermore, on-site food is now offered by 10% of wineries, an increase from 4% in 2010.

It is also claimed that the general attitude towards wine tourism is changing, with half of the wine-growers convinced that wine tourism can boost their sales and increase customer loyalty by developing long-term relationships. However, this is controversial, with others denying such a trend.

Typically, wine tourists in the Côte d'Or have three main motivations for visiting the region (Charters, 2015): the discovery of the wines and vineyards (44%); the discovery of gastronomy and regional produce (27%); and the discovery of the cultural heritage (21%). Clearly, wine tourists come to Burgundy without a sole focus on the region's wines. Meanwhile, according to a survey conducted by the Beaujolais Interprofessional Committee (2010) at three local wineries, the average wine tourists in Beaujolais are around 45 years old, 65% are foreigners, 35% are executives, 21% are retired and 18.5% are employees. Their key motivations are the following: 40% come to taste wines, 37% to enjoy food and wine and 28% to visit cellars and wine-growers. For over 50% of them, wine has been a decisive factor to visit this region. They spend on average €70.50 on wine, which is low compared to the average in French wine regions (€104). During their stay, they tend to do some wine tasting, eat in restaurants, and buy gourmet food products. What is especially interesting from this survey is that their perception of Beaujolais wines has improved after a stay in this region.

THE PROBLEM: TOO MUCH COMPLEXITY

There are clear strengths for wine tourism in Burgundy. The wines have been famous for hundreds of years. There is a varied choice of accommodation from top-class hotels to bed and breakfast, and self-catering. Both the Côte d'Or and Beaujolais are easily accessible through the A6 motorway, with links to Paris and its airports and south to Lyon and beyond with 75 million travellers per year (Burgundy Tourism.com, 2014). It is also "en route" for tourists travelling around many places in France. Additionally, gastronomy is a clear draw card, from the "Bistrots du Beaujolais" which promote local wines to Michelin-starred restaurants.

However, there are clear weaknesses in the wine tourism offer throughout Burgundy, and much unfulfilled potential, including the following:

- *Elite Wines & Complex Appellation System* - Burgundy wines generally suffer from an elitist image of only being accessible to a small minority of wealthy people. A lot of wine tourists in Burgundy, especially foreigners, find the whole experience daunting and intimidating (Hyde, 2013). Furthermore, the complex system of wine appellations is difficult to understand for the average wine tourist.
- *Difficulty Obtaining a Tasting Appointment* - Advance booking for wine-tasting is almost always compulsory. This can be daunting, especially if the phone-call has to be made in French, as the older

wine-growers often do not have a good command of English. If visitors arrive unexpectedly to the domaine, there may be no-one to deal with them. This is, of course, a problem for wine tourism worldwide, but the linguistic problems for non-Francophones exacerbates it, as many potential visitors fear phoning ahead and only turn up on speculation. It is also a challenge to find a winery open on the week-end, especially on Sunday. This is due to the small size of the estates, traditionally managed and owned by one single family.

- *Lack of Coordination Between Tourism Organizations* – Though France has an excellent system of Tourist Information Offices in most major cities, co-ordination between offices is limited. Tourism offices in major towns will often not give information about regional tourism, but will only focus on their own town. Furthermore, there is often variation in information provided between differing levels of regional, departmental, and city government, with an apparent lack of coherent planning.

- *Need for a Mass-Market Wine Tourism Website* – though there are a variety of websites that advertise wine tourism in Burgundy, the plethora of these can be confusing to the average tourist. There is a need for a mass-market wine-tourism site that is user friendly and allows tourists to contact wineries, schedule appointments, and find links to individual winery websites.

- *Lack of Collaboration Amongst Wineries* - there is an individualistic approach to business which makes the cooperation necessary for effective wine tourism more difficult. Typically domaines are reluctant to cooperate with each other to promote jointly their territorial wine tourism brand (Charters and Spielman, 2014). Equally, they often seem reluctant to assist visitors who are looking for a wider experience by suggesting other attractions to them (Hyde, 2013).

There is, consequently, a sense that in Burgundy, despite the reputation of its wines and its appeal as a general tourist destination, the wine tourism offer is not as well developed as it might be.

Specific Tourism Challenges in the Côte d'Or

There are, in addition, some specific problems surrounding the image of the Côte d'Or. As well as the issue of aristocratic wines sold at a high price, there is a perception that quality can be quite varied, especially based on vintage variation. The architecture of the domains can also be intimidating, because they are usually old stone-houses with private courtyards, and not obviously open to the street. There is often no clear signage of the name of the domaine with opening hours listed. As a result, wine tourists might feel unwelcome and unwanted. It has been said that at the overwhelming majority of wine providers the word welcome is not uttered when someone arrives (Charters, 2015). This may be in part a matter of cultural difference, but it merely reinforces the image of exclusivity and intimidation.

Specific Tourism Challenges in Beaujolais

Whilst the wines are very affordable in Beaujolais, nevertheless the sub-region does suffer from the image given by Beaujolais Nouveau of cheap, poor-quality, and unfashionable wines. For example, Delphine d'Harcourt from "Château de Montmelas" reported that when she is at a wine fair in France or abroad, she tends to make people taste her wines before telling them from which region they come from. She promotes the brand "Marquis de Montmelas", instead of Beaujolais Villages, which can be more difficult to sell. Bruno Metge-Toppin has a similar approach in that he sells his wines as "Burgundy" rather than under the "Beaujolais" appellation. In this way he gets a higher price for them.

Other specific challenges to the Beaujolais region include a lack of top-class accommodation, no unified website advertising wine tourism in Beaujolais, and no regional organization to develop and promote wine tourism. This means there is a lack of research on the profile of wine tourists to Beaujolais, including why they come and their needs. Nobody is taking on responsibility for establishing this information.

THE SOLUTION: BUILDING ON THE STRENGTHS OF BURGUNDY

Whilst some aspects of the failure to develop an effective wine tourism in Burgundy seem specific to the different sub-regions, there are also some common failings which appear more institutional and need a collective approach, if they are too be remedied. Fortunately the region seems to have some strengths that it could further harness to overcome the problems, as well as some opportunities which may allow it to develop positively. These are described below:

Crucially, the entities in the region need to focus on planning and coordination, on image building using events and attractions, and on harnessing the dynamism of some key individuals who are attempting to make the region work effectively.

Planning and Coordination: Working Together to Build a Regional Wine Tourism Brand

To be successful, both sub-regions need to develop and market a defined and strong territorial identity. It appears that personality is necessary. Competition is fierce and tourists have the choice between numerous destinations within France. At a practical level, clearly signposted and marketed wine routes or trails, which cross political boundaries, are essential in facilitating wine tourism. More carefully targeted wine events, especially for lower involvement consumers, regional stars, and a diverse product would also contribute to ensuring that tourists become repeat visitors.

While the reputation of the Côte d'Or can be elitist, making some feel intimidated, that same reputation is a powerful magnet to some visitors. This reputation therefore needs to be carefully managed for the benefit of the entire sub-region. As Caroline Licati from Parigot and Richard, a crémant producer in Savigny-lès-Beaune, stated:

> *"Wine-growers in Burgundy often have an attitude problem,*
> *especially if they sell all their wines through exports. The young*
> *generation is even more arrogant than their parents. They do not*
> *see the point of wine tourism and marketing is a dirty word for a*
> *lot of them".*

That is why when she receives foreign wine tourists, she often takes them to other, bigger places like "Château de Pommard" or "Château de Meursault", to offer a full sense of the region.

Beaujolais is geographically accessible through its close proximity to Lyon. Furthermore, Lyon International airport is also quite close, and this offers a clear opportunity. Three and a half million tourists stay in Lyon every year, but only a small minority visit Beaujolais during their stay. The city of Lyon has started to promote Beaujolais as an attractive destination for business travellers. Beaujolais could be marketed at the "secret garden of Lyon". A clear regional plan with a strategy and a budget is necessary and collaboration between all stakeholders is vital.

It is clear that there is a great need for collaboration at all levels. There should be a close collaboration between the tourism and wine industries, between public and private organizations, between the tourism actors and all wineries of a destination, and also between the wineries themselves, which, too often, operate in too individualistic a manner. The future of wine tourism for both regions could be to linking more wine and food with wine and culture. Their point of differences should be clearly marketed.

Expanding and Promoting Regional Attractions

Both sub-regions have a number of attractions which must be accepted as integrated into the wine tourism offer. As noted above, these attractions need to work together, and be part of a coordinated offer.

The Imaginarium in Cote d'Or - is a purpose built showcase dedicated entirely to sparkling wines, based in Nuits-Saint-Georges. It is designed to entertain children as well as adults and exhibits are in English, French, and German. All sparkling wines from around the world are represented. It is located at the Louis Bouillot production site, conveniently situated near a motorway exit. The site is big, 1,200 square meters, and takes the visitor through three separate stages: an interactive journey of discovery, a film show, and a tasting. This is probably one of the best wine experiences for children in Côte d'Or.

Négociant Showrooms in Cote d'Or - like Jadot and Drouhin based in Beaune who have opened their production facilities or their historical wine cellars for visits. Domaine Drouhin has decided to opt for a high-class tour of their cellars and the tasting of six wines, for wine connoisseurs. Prices range from 38€ up to 250€, which includes a special dinner. The price allows them to focus only on high-involvement consumers, and the visit is highly educative. The tour takes place in the old buildings in which wines were made, with stories about the history of the house and the family, including the privations they went through during the Second World War, as well as wine production. However, there is a need for other less demanding, more experience-focused visits suitable for lower-involvement consumers.

Small Village in Burgundy

Hameau Duboeuf in Beaujolais – is a major tourist attract with more than 100,000 visitors a year. Organized as a fun wine village experience for the whole family, it is a gateway to Beaujolais and especially Beaujolais wines. After visitors have spent a day in his park, the owner sends them to discover the Crus du Beaujolais and explore the region.

Château de la Chaize in Beaujolais - with its classified château, gardens and cellars has become a key attraction for French and foreign tourists. They sell 50% of their wines through exports, the rest through hotels and restaurants, but wine tourism is important for them, as a public relations tool. As Marquise Caroline de Roussy de Sales said:

> *"If I try to summarize what wine tourism is, it is a good way to develop and maintain relationships with business and leisure*

customers. It is a different relation with wine, as visitors can learn what and who is behind the wines we produce. When they come to the château and I introduce our wines to our American importers, they personify our wines and they won't forget me afterwards! Visitors, in general, are interested in knowing how wine is produced; they are looking for authenticity and a personal relationship with the wine-grower or the domain owner".

Château de Montmelas - the ancient residence of the noble family from Beaujeu, the "Château de Montmelas" has been in the same family since 1566. Countess Dephine d'Harcourt is a very active lady; on top of bringing up her four children, she is the sales manager for the estate. Throughout the year, she organizes special events at the chateau, linking wine to food or to gardening and also proposes a tasting at 11 a.m. every Saturday. She also manages a gîte on site which allows her to sell wines to a captive audience. The Countess strongly believes in wine tourism. Thanks to her initiative and dynamism, she manages to sell a very high percentage of wines through the cellar door.

Promoting Regional Events

Some key wine events and festivals attract thousands of business and leisure tourists every year to Burgundy. These events need to be promoted more broadly, and embraced by all entities as a source of pride and common communication. Following are some examples:

The Festival of la Saint-Vincent Tournante celebrates each January the Patron Saint of wine-growers. This happens throughout the region, but it is a major spectacle in the Côte d'Or, taking place in a different village each year (Demossier, 2010), and attracting many thousands of tourists as spectators to a key event in the life of the growers.

The Hospices de Beaune Wine Auction in November remains a key event for wine importers, when the new vintage is evaluated and its worth is established via a major and historic wine auction. Non-professional visitors are attracted by the historic ritual and contemporary glamour of this event.

Les Grands Jours de Bourgogne is a major event takes place every other year in Beaune, where importers, journalists, wine writers, and various influencers are invited for a week in spring to taste the wines of the different sub-regions of Burgundy. Professionals are also tourists, and probably spend more per capita during visits than non-professionals. They should thus not be overlooked.

Fête des Crus du Beaujolais is held every year around mid-April this festival takes place in a different village of the ten crus of the region. Over the week-end, wine tasting of all the crus takes place. The event seems to attract mostly French people and especially locals, but it may need to expand its reach to attract a wider audience.

Tour de France is held during the "Tour de France", it is possible to collaborate with organizers to encourage tourists to visit wineries. For example, in a more recent year, organizers passed out coupons which allowed visitors a free tasting at various places in Beaujolais. It was a major success, as it allowed thousands of visitors to discover the wines of the region. This type of collaboration with an online national event is a great opportunity for Burgundy to continue in the future, and expand to other sub-regions.

Promoting Positive Individual Tourism Efforts

Wine tourism in Côte d'Or is still relatively undeveloped, however, there are some key successful private initiatives that are noteworthy. For example, some wine-merchants are developing beyond merely offering wines for sale, but adding wine education, visits to the vineyards, accommodation, and food on-site. A good example is the role of Olivier Leflaive in Puligny Montrachet. He set up his own négociant business trading in wine, and then decided to create the "Table d'Olivier Leflaive" in the village with a menu paired with his wines. The servers in the restaurant describe the wines, explain the vineyards, and provide maps for diners to consult. Olivier is considered to be a wine tourism pioneer, because he was the first to offer this type of attraction in Burgundy. The impact has been very positive in that the restaurant attracts many visitors to the village, and they become brand ambassadors for the wines and the place.

In Beaujolais, it is generally accepted that the local people are very warm and hospitable. This should be underlined as part of the offer. Within the wine industry there is a generation of young wine-growers who have considerably improved the quality of their wines and a lot of them practice organic viticulture. Again, this could be made into a strength, tied to the local environment and marketed especially to younger and urban, environmentally-focused tourists. Again, the role of Georges Duboeuf is most significant here in that he created a place of education and entertainment which introduces visitors of all ages and all levels of wine knowledge to discover the wines of Beaujolais.

What is most important about encouraging these individual initiatives is that, when they work for the benefit of an entire village or area, they can help to breakdown the isolation that exists behind the wine tourism offer, and act as a beacon of best practice for all the providers.

Summary

Wine tourism in Côte d'Or seems relatively under-developed, but things are improving slightly. Its success lies in collaboration between public and private actors, between the wine and the tourism industries, between the wine-growers themselves. There is also a need for a regional plan for developing and supporting wine tourism with a budget, a clear target market and a clear positioning.

In this chapter, we have tried to identity some key factors for a successful development of wine tourism from a regional point of view. In a region, all actors involved in wine tourism activities should adopt a certain number of best practices which would enable them to offer a truly unique experience to tourists The first key success factor is knowledge of the market and the competitors. Beaujolais clearly lacks suffers from a lack of knowledge in the profile of its wine tourists. More market research should be carried on: it would help them meet visitor expectations and understand their motivations, thus allowing the actors involved to refine their wine tourism offer.

At a winery level, providing a truly memorable experience is essential. Research has shown that tourists aspire to enjoy a total experience. It is the quality of the service they experience, and not the quality of the wine they taste, that matters to most visitors. Wineries who want to develop their wine tourism activity should invest in staff aware of the need for quality service and in their training. Some small wineries in Beaujolais and in Côte d'Or mainly see wine tourism as an alternative distribution channel and only start a wine tourism business with that goal in mind. To be successful, they should shift their approach from just selling their wines to marketing their unique wine tourism experience in order to build long-term relationships with their customers and foster brand loyalty.

As a conclusion for both regions: Beaujolais seems to have an even bigger wine tourism potential than Côte d'Or, the difference being that wine-growers have more difficulties selling their wines, especially Beaujolais Villages, so they are more open to wine tourism. The same situation is to be found in other French regions, like Languedoc-Roussillon or the Loire Valley. Wine tourism may help remedy the poor image of a region, such as the damage caused by the Beaujolais Nouveau in Beaujolais, and can lead tourists to discover a region and then its wines. In any event, Burgundy, already one of the world's great wine producing areas, has great potential to expand its position as a key wine tourism destination.

DISCUSSION QUESTIONS

1. How can history, terroirs and gastronomy be used as wine tourism leverage?
2. What could we adapt from the New World to these two sub-regions? And why? What practices could not be implemented in the Old World and why not?
3. Is the development of wine tourism a viable option for all wine regions and wineries?
4. What additional opportunities are there for Burgundy to expand its wine tourism offering?

Chapter Seven

Pink Wine & Movie Stars: How the Provence Wine Trail Was Established

Coralie Haller, Sébastien Bede, Michel Couderc, & Francois Millo
University of Strasbourg and Provence Wine Council

On a beautiful spring day in April 2007 in Provence, when most tourists were enjoying the sunshine, the rich cultural and natural heritage, and the cuisine and wine of the region, the members of several Provence wine tourism organizations were in a meeting. There was a sense of frustration in the room, as they discussed different methods to enhance the tourist appeal of Provence vineyards. Finally one member spoke up, expressing the viewpoint of many:

> *"C'est quand même incroyable qu'un département comme le Var, tellement touristique et avec énormément d'offres de produits, n'arrive pas à communiquer de façon cohérente et coordonnée vers les clients." (Translation: "It's quite incredible that the Var area, with so many tourists and such a huge product offer, cannot communicate to its customers in a consistent and coordinated way.")*

The statement highlighted not only the poor visibility of wine tourism offerings in Provence, but also the increasing confusion around the definition of wine tourism in the region. This was because there were a multitude of different and competing wine activities including: wine tasting with industry experts, visits to wineries and tasting rooms, nature walks through the vineyards, meals at country inns and bed & breakfast establishments, and many other wine related events.

Unfortunately, there was very little communication or coordination between the various stakeholders in the Provence wine and tourism industry, and this was causing great confusion, overlapping activities and a waste of resources. Therefore, the members of the various organizations were meeting to try to build a coherent strategy and create the Provence Wine Trail.

OVERVIEW OF THE FRENCH WINE INDUSTRY

The French wine industry is important both economically and commercially. Worth around 9.5 billion euros in 2012, compared to 5.5 billion euros in 2003, wine generates the second largest export revenue in the French economy, just behind aeronautics at €20 billion (FEVS, 2013). With a production of 46.2 million hectolitres of wine in 2014, France remains one of the three largest wine-producing countries, along with Italy and Spain (OIV, 2014). 30% of the wine produced in France is exported, mainly to European countries, with sales of 13.7 million hectolitres representing an export turnover of 7.6 billion euros in 2013 (Vin et Société, French customs data, 2014).

Although France is 'only' the second largest wine drinking nation in the world, after the United States, consuming 28.1 million hectolitres (OIV, 2014), wine remains a national drink. In this matter, France has endorsed a special protection for its *"wine and viticultural terroirs"* by incorporating them into France's cultural, gastronomic and landscape heritage (OIV, 2014). Vines occupy 3% of total land area in France, which makes it the world's second largest vineyard area after Spain, with 755 000 hectares of vines grown in the country in 2013 (France Agrimer, 2013).

France has eight primary wine-producing regions: *Alsace, Bordeaux, Burgundy, Champagne, Languedoc-Roussillon, Loire Valley, Provence* and *Rhone Valley*, and smaller wine-producing regions that include *Corsica, Jura, Moselle Wine, Savoye-Bugey* and the *South-West* (Johnson & Robinson, 2013). France's wine regions are renowned for their specific grape varietals like the Syrah grape in the Rhone valley, the Cabernet Sauvignon and Merlot grape in Bordeaux, the Sauvignon Blanc grape in Bordeaux and the Loire valley, the Chardonnay grape in Champagne and Burgundy, and the Pinot in Burgundy. The country produces all kinds of wine, from white to red and rosé wines, dry to semi sweet and sweet, as well as sparkling wines and fortified wines.

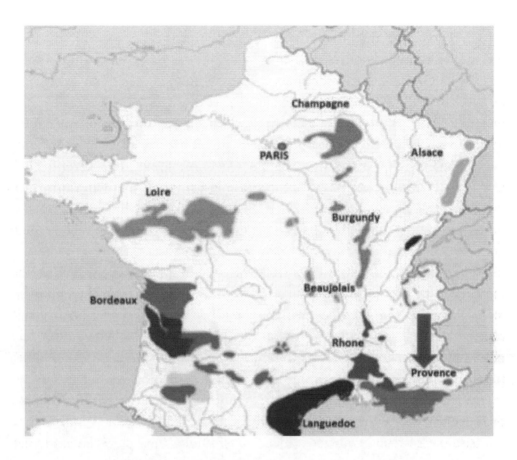

Map Showing Location of Provence in France

In France, the *Appellation d'Origine Protégé* (AOP) and *Indication d'Origine Protégée* (IGP) system was developed to protect the Designation of Origin (DO) for wine, with strict laws concerning winemaking and production in relation to the 'terroir'. The terroir is based on natural characteristics such as soil, rock density, altitude, slope, sun orientation and microclimate, including rain, winds, humidity and temperature variations, together with the grape varietals and wine growing and production methods. In the last ten years, AOP wines represented 46% of French wine production on average, while IGP accounted for 30% (France Agrimer, 2013).

Wine Tourism in France

In France, vineyards are a key development factor in the French tourism industry, because the revival of wine tourism in the 1980s was one of the solutions developed by farmers to address economic challenges. More generally, wine tourism gives winemakers the possibility to diversify their income with direct sales to tourists (Mallon, 1996), raise the profile of their vineyard and promote local heritage and know-how.

Acknowledging the need to make the wine world more accessible and understandable to a wider audience, winemakers and tourism organizations have developed a range of activities and attractions based on the theme of wine in an attempt to encourage visitors to cellars and wineries. Over the last five years, there has been a steady increase in the number of vineyards open to the public, the number of wine trails and the level of tourist infrastructures. In some regions, the majority of properties open to the public are small businesses. The winery owners themselves usually provide the services on offer. Because of their small size and fragmented nature, the marketing and promotion of wine tourism for these smaller cellars is limited to displaying a sign outside their building, participating in a winemaker's charter, listing in a wine guide (the most popular is the Hachette guide), and providing more or less detailed information on social media and other websites.

Wine tourism is not limited to the smaller cellars, however. In a number of French wine-producing regions, several larger producers have also made tourism one of their major activities, transforming their wineries into very professional, commercially run and popular tourist attractions.

| # of Wineries in France | 85,000 |
| # of Wineries in Provence | 600 |

At present, however, the wine tourism industry in France remains highly segmented. In fact, in 2010, around 10,000 of France's wine cellars counted over 10 million visitors (Atout France, 2015). Of these, 39% were from other countries, including Belgium, the UK, the Netherlands, Germany and the US. It should be noted that France boasts 31 museums and themed wine sites, which receive more than a million visitors a year.

In 2009, the French tourism organization, Atout France, created an award called *Vignobles &* DECOUVERTES to certify French wine tourism destinations. The award is for 3 years, and is administered by the Ministry of Tourism and Agriculture after recommendations by the Upper Council of Wine tourism. Currently 36 wine tourism destinations in France have received this award.

Although wine tourism is said to play a *"growing and fairly significant role in France"* (Choisy, 1996), it is less important to the wine industry than in many New World wine regions. Despite a growing awareness of the potential of wine tourism in France with regard to diversification of farm economies and regional marketing, researchers (Choisy, 1996; Desplats, 1996; Mallon, 1996) have noted a generally low participation rate by wine producers in the tourism industry. However, wine has long been a significant component in the promotion of France as a tourist destination in terms of both image and opportunities to experience French cuisine and wines.

OVERVIEW OF THE PROVENCE WINE REGION

The first vines and wine growing culture were introduced to Provence around 600 BC, with the founding of a colony in Marseilles by the Greeks (Palanque, 1990). The wines made in those days had a light colour, similar to rosé wines, because maceration of juices with grape skins was either unknown or only practiced on a limited basis. Provence is therefore the oldest wine-growing region in France, and the first wines to be made were rosés.

Map of Provence Wine Region

Located between the Mediterranean and the Alps, Provence's vineyards extend from West to East over approximately 200 kms (120 miles), primarily in the French departments of the Var, Bouches-du-Rhône and

Alpes-Maritimes. There are three main appellations in the region, representing 96% of the volume of wines with Provence appellations:

1) The *Côtes de Provence* Appellation is located in the eastern region of Provence. It includes the recently recognized denominations of the Côtes de Provence Sainte-Victoire, Côtes de Provence Fréjus, Côtes de Provence La Londe, and Côte de Provence Pierrefeu terroirs.

2) The *Coteaux d'Aix-en-Provence* Appellation is located in the western region of Provence near the town of Aix-en-Provence.

3) The *Coteaux Varois en Provence* Appellation is located in the central part of Provence, clustered around the village of Brignoles.

There are approximately 600 winegrowers (540 independent growers and 60 cooperative wineries) and 40 wine merchants producing 1.3 million hectolitres of wine, the equivalent of 174 million bottles produced on 27,000 hectares of land.

The Importance of Rosé – Pink Wine

The region of Provence is quite important to France because it is the largest producer of AOC rosé wines, representing 40% of domestic production and 5.6% of the world's total rosé wine production. With 90% of the local wine produced as rosé, this represents an average annual output of 150 million bottles of AOC rosé (CIVP, 2014). Making rosé wines has been an integral part of the 'art of living' in the region for generations as Provence's climate, soils and grape varieties are all perfectly suited to rosé winemaking. Rosé wine consumption has increased year on year since 1990, and in 2014 it accounted for 30% of total wine consumption in France (CIVP, 2014). Consumers immediately associate the region's winemakers with rosé wines.

Rosé wine has become trendy in France, as almost 9 out of every 10 wine consumers say they drink rosé, representing a total estimated market of 36 million rosé wine consumers (CIVP, 2014; Wine Intelligence, 2014). However, rosé is more than just a passing trend. It has become an integral part of French society, associated with new consumer trends and lifestyles:

e.g., less structured meals, the increasing popularity of ethnic cuisine from around the world, greater simplicity, new encounters, good times, and instant gratification. When drinking rosé, the consumer discovers a different approach to wine: easier access and greater freedom without all the constraints and traditional formalities.

The newfound public for rosé wines has extended well beyond France's borders and won over wine drinkers worldwide. In the past 10 years there has been a 15% increase in rosé consumption globally, with 22.3 million hectolitres consumed in 2012 versus 19 million in 2002 (Observatoire du Rosé CIVP, 2014).

Drinkers have demonstrated a preference for dry, light-colored rosés, and Provence's winemakers are recognized experts in this type of wine. Provence's rosés are thus well positioned to win over new customers abroad. Consumers in the United States have proven enthusiastic, for example, as imports of Provence rosés increased threefold between 2003 and 2013 (CIVP, 2014). Although Provence is historically recognized as a producer of pale, fruity, and full-bodied rosé wine, the region's wineries also produce powerful and well-structured red wines that can age several years, as well as delicate whites renowned for their lightness and subtlety.

Women Drinking Rose Wine

Rosé wine remains a dynamic market with constantly growing demand, and has become an important element of the global wine offer in terms of production, consumption, economy and trade. As a consequence, the Provence region is one of the rare wine-producing regions not to suffer from what is seen as a crisis in the other vineyards.

THE PROBLEM: LACK OF CLEAR STRUCTURE AND ACCOUNTABILITY FOR WINE TOURISM

Provence has been known for centuries as a great destination where visitors can see fields of lavender, quaint villages, seaside resorts, and drink rosé wine. It has also attracted many celebrities and movie stars, including the recent arrival of Brad Pitt and Angelina Jolie who purchased a winery and vineyard (Camuto, 2014). Despite all of these positive aspects, in 2007, wine tourism in Provence was a confused tangle, with lack of a clear vision, structure, and accountability. This was due to several problematic issues:

1) Fragmented wine organizations - Part of the issue had to do with the many complex and fragmented professional wine organization operating in Provence. These include wine councils (production and trade for DO wine products); appellation trade unions for DO wine products, federations of independent winegrowers and cooperative cellars, terroir associations and public administrations. They were all encouraged to exchange economic and declarative information, knowledge, knowhow, and innovations, and to pass on information to their members. Unfortunately, this was not always happening in a consistent fashion.

2) Unclear geographical boundaries - A second major issue had to do with the fact that there was not a clear understanding of the geographically boundaries of what is called the "Provence wine region". From an administrative perspective, the Provence region is part of the PACA region (Provence Alps French Riviera region). However from an appellation perspective, Provence is organized around the Provence Wine Council (CIVP), a regional professional wine body which represents both production and trade professionals for the three major appellations: Côtes de Provence, Coteaux d'Aix-en-Provence, and Coteaux Varois en Provence.

3) No clear authority amongst tourism organizations – A third issue had to do with the lack of a single authority that has exclusive rights or complete jurisdiction over tourism. Instead, numerous organizations are active in tourism at national, regional, and local level. For administrative purposes, France is divided into 22 regions that, in turn, are subdivided into 96 (plus 4 overseas) departments, each with their own administration. The jurisdiction of tourism is embedded in each administrative layer: at regional level it is governed by the *Comites Regionaux du Tourisme* (CRT), at departmental level, the *Comités Departementaux du Tourisme* (CDT), and at local level, the *Offices du Tourisme* (OT).

Regions and departments provide funding for tourism committees (*comité du tourisme*), their main technical tool for the planning and management of tourism. Tourism committees include representatives from the regional council (*conseil regional*), tourism professionals, consular organizations, tourism associations, and tourist resorts. Regions and departments may entrust all or part of the implementation of their tourist development policy to these structures, but this varies from one region to another.

Therefore, the French tourism industry is relatively disorganized, and the competencies of many of the different tourism entities overlap. This fragmentation of power has led to significant changes in the ways that central ministries, local government, and regional authorities manage the tourism sector, raising issues with respect to costs, service quality, and governance conflicts.

4) Imbalance of tourist amenities - A fourth issue that was causing some conflict had to do with the location of tourist facilities. Most of the facilities are concentrated around Marseille and Nice. For example, there are two regional tourism offices, one in Nice, created in 1942, covering the "French Riviera", and another in Marseille, created in 1987 and covering the "Provence-Alps-French Riviera" region. Though the two offices signed a cooperation agreement in 1987, it is still challenging for some of the regions to receive enough attention and coverage. A case in point is the Provence Alps French Riviera (PACA) region, which is subdivided into four departmental tourism entities, namely, the Comité Départemental du Tourisme for the Var, the Bouches-du-Rhône, the Vaucluse, and the Hautes-Alpes. For them, having only two regional tourism centers does not provide enough coverage and support.

5) Muddled marketing messages– Because of the lack of clear roles and accountability between the various parties, there was an issue of muddled marketing messages. Often they sent out different messages and brands, scrambling the destination's image and confusing tourists. In establishing Provence as a wine tourism destination, conflicts emerged, as often happens when many groups with different goals are involved and the interactions between them are difficult to identify, especially when no composite picture is available.

THE SOLUTION: DEVELOPMENT OF THE PROVENCE WINE TRAIL

The situation came to a head in 2007 when the Var Chamber of Agriculture became frustrated over the fact that they could not promote their local produce as well as they desired. The lack of clear roles and accountability in the region made it very difficult for them to achieve the attention they needed for their agriculture products. However, instead of merely complaining, they decided to take a leadership role and bring together several key stakeholder groups, including national councils, tourism, agriculture and wine organizations, and two regional banks. They were determined to find a solution to the many issues plaguing tourism in Provence.

After several meetings, the idea of developing the Provence Wine Trail slowly emerged. They realized that a common vision was to develop a unified tourism concept shared by all stakeholders that would encourage consumer and producer interaction. The idea of using wine as connecting point, via the Provence Wine Trail, would allow all parts of Provence to be unified with this strategy. With this in mind, they set about developing a series of processes and tactics to make to implement the strategy. Altogether, it took four years to design and implement.

Formation of Formal Stakeholder Group to Develop the Trail

The original stakeholders who attended the initial meetings with the Var Chamber of Agriculture were invited to be part of the formal stakeholder group, but others were invited to join as the project grew in size. To manage the project, the PACA regional Chamber of Agriculture took on the role of project coordinator, and the Var district was appointed as project leader. Following is a list of the stakeholders who helped to develop and implement the strategy:
- **State organizations**: PACA Council, and the *Conseil général* of the Var, Bouches du Rhône, and Alpes Maritime
- **Tourism organizations**: *Agence du tourisme* of the Var and the Bouches du Rhône

- **Agricultural organizations**: Var Chamber of Agriculture
- **Wine entities**: Provence Wine and Intervins SE council; and trade unions for all local DO wine products (over 25 of them): the AOP of Bandol, Baux-de-Provence, Bellet, Cassis, Baux-de-Provence, Palette, Coteaux d'Aix-en-Provence, Coteaux de Pierrevert, Coteaux Varois en Provence, Côte de Provence Fréjus, Côtes de Provence, Côtes de Provence La Londe, and Côtes de Provence Sainte-Victoire, and the IGP Alpes de Haute Provence, the VDP of Sainte-Baume, Alpilles, Bouches-du-Rhône, Coteaux du Verdon, Maures, *Portes de la Méditerranée*, and the Var, the *Syndicat des vignerons du Var*, the *Syndicat des vins de Pays des Bouches du Rhône*, and the Crus classés
- **Wine organizations:** Regional and departmental Federation of Independent Wine Makers, Federation of Cooperative Cellars
- **Banks:** Credit Agricole Provence Côte d'Azur and Credit Agricole Alpes/Provence*

Financial Funding for the Provence Wine Trail

The next step was to identify funding to help make the trail become a reality. After some discussion, the project was funded with 430,000€ from PACA Regional Council, representing 41% of the total budget of 1.06 million euros. The other financial partners were the Var Chamber of Agriculture, the Provence Wine council (CIVP), and a regional bank.

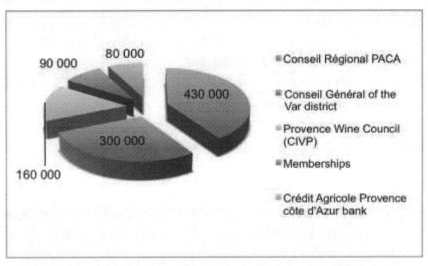

Provence Wine Trail financial resources (2007-2011)

Fleshing Out the Vision and Scope of the Project

The stakeholders also spent much timing discussing how they envisioned the Provence Wine Trail working in the future. They agreed that their aim was to help tourists to discover and visit the Provence vineyards through the application of various tools, such as clear signage along wine routes, brochures, a unified logo and advertisements, and a supporting website.

They also agreed to include all of Provence. Originally the Wine Trail was supposed to exclude the "French Riviera" players. However, the organization in charge of wine promotion and tourism along the Rivera did not have a tourist information center. Instead, it was a professional wine organization of wine producers and dealers, which made it possible for the CIVP to overcome any administrative barriers and include them. Thus the Provence Wine Trail was able to include wineries from this area (French Rivera and Alpes-Maritimes) to promote wine tourism in the southeast of France.

Development of a Regional Brand with CIVP in Charge of Communication

Another important part of the project was development of a regional brand. After much work and discussion, the stakeholders developed a logo of a stylized cluster of grapes on a blue background with the words "Route des Vins de Provence." All writing was in white with a small pink grape and pink "o" to highlight the importance of pink rose wine.

It was decided that the Provence Wine Council (CIVP) would be in charge of coordinating and organizing wine and tourism communication campaigns, notably the promotion of authentic wine products and raising consumer awareness.

Development of Marketing Materials and Signage

Once the brand was agreed upon, the stakeholders worked on developing marketing materials, including laying out the wine trail, and creating wine maps, brochures, advertisements, and other brand-related merchandise to promote the region. They also began the process of installing the signs along major roadways so tourists could easily find the wineries.

Provence Wine Trail Marketing Materials with Logo

Development of Dedicated Wine Tourism Website

Originally devised as a simple concept of a "physical drive" along the wine tourism trail, the project gradually evolved to include a "virtual drive" along the wine trail through the development of a dedicated website: http://routedesvinsdeprovence.com.

The website allows tourist to prepare for their visit by providing precise descriptions of each vineyard and cellar, plus simple tips and tools to plan their stay. Tourists can also create their own itinerary and share their experiences on social networks linked to the website. Other information on the site includes articles produced by a journalist and a photographer. These feature the history, viticultural methods, diverse products, and unique characteristics of each participating winery.

Invitation to Wineries to Sign Charter

Another aspect of the project was to invite all of the wineries along the Provence Wine Trail to sign a Wine Trails Welcome Charter in which they agreed to welcome and inform all visitors, whether experts or novices. This charter also set out the specific criteria that wineries have to comply with in terms of reception arrangements, organization of visits, childcare, etc.

Training for Wineries on Tourism Hospitality

To perform more effectively, winemakers were invited to special training sessions organized by the project steering committee, where they learned to optimize their cellar management, and how to organize visits to their winery and wine tastings.

Wine Trail Annual Membership Fee

To demonstrate their commitment to tourism and being a partner, wineries were asked to pay an annual membership fee to join the Provence Wine Trail network. The fee is 200 euros per year, and many of the wineries gladly paid the amount to be part of the new tourism strategy.

On-going Marketing, Special Events, and Promotion of the Provence Wine Trail

Today, the Provence Wine Trail continues to operate with great success, though there are still some challenges and issues that the CIVP and other stakeholder members must tackle. As Eric Altero, the project coordinator from the Var Chamber of Agriculture, stated:

> *"Il n'est pas facile de faire travailler ensemble des domaines privés et des coopératives, le tout articulé avec plusieurs appellations et plusieurs départements". (Translation: "It's not easy to get independent wine growers and cooperative cellars to work together, coordinating everything between several appellations and several departments.")*

Examples of Provence Wine Trail signs

However, despite everything, the Provence Wine Trail has been fully operational since the summer of 2011. The CIVP has taken a leadership role in coordinating marketing and promotion activities, as well as helping to coordinate special events. Some of these events include the "Vigneron's Cup", a frigate around the wine making island of Porquerolles, "Art et Vin", which presents diverse forms of art in the vineyards during the summer, and the "Ballades Groumandes en Terroir Pierrefeu", where people walk and taste wine directly in the vineyards.

RESULTS AND WHY THE PROVENCE WINE TRAIL IS A TOURISM BEST PRACTICE

There is no doubt that the Provence Wine Trail project has been a huge success and is considered a wine tourism best practice in France. In a region where there were many different wine, agriculture, and tourism organizations that were not working together and were accidently creating chaos and confusion for tourists, these various parties came together and created a unified brand and wine trail that are now known all over the world. By working collaboratively they have helped to propel the Provence brand into a new realm, and now tourists around the world come to the region not only to taste "pink wine" and view movie stars, but to explore the unique cultural heritage, history, and other agriculture products of this region.

Statistics also testify to the success of the Provence Wine Trail. To date, 430 of the 600 wineries in Provence have signed the Wine Trails charter and are committed to welcoming tourists with open arms. According to the Provence-Alpes-Côte d'Azur (PACA) Tourist Board (2014), Provence is now the second most visited region in France after Paris, with 31 million visitors in 2013. Furthermore, they report that tourism in the region generated €14 billion in revenues and accounted for 11% GDP in 2013. There has also been a pronounced increase in sales of Provence wines in the past several years, up to 40% in volume, all retail included (CIVP, 2014). Though this cannot all be attributed to the wine trail, the positive impact of a unified Provence wine brand has certainly impacted these numbers in a positive way.

Other best practice results can be linked to the mode of governance, tools, development of the Provence brand, and agreement on geographical boundaries.

Wine Trail Governance

The Provence Wine Trail project's success is closely linked to its mode of governance. In a fast-paced wine tourism environment, the stakeholders chose to share power and resources rather than adopt a centralized governance system through a single organization. By paying close attention to everyone's goals, the stakeholders identified and combined their assets. It required human resources to coordinate and lead the project, and financial means to support it through the 4-year process.

Moreover, the stakeholders all shared the project leadership as intermediaries with different roles: as a public-private actor, the PACA regional Chamber of Agriculture is the main financial contributor and therefore plans, organizes and manages the resources. As a public body, the Var Chamber of Agriculture is responsible for getting the stakeholders to cooperate. The Provence Wine Council (CIVP) is in charge of coordinating and organizing wine and tourism communication through the promotion of the Provence Wine Trail. With a small number of stakeholders involved in the project, governance by just one agent could result in the inefficient use of resources. Sharing the governance between three legitimate stakeholders appears to deliver effective results due to accountability, involvement by a subset of network members, and consensus in terms of goals and outcomes.

The Tools: Website, Signboards and Road Legislation

Interestingly, the use of Internet technology made it possible to successfully initiate and coordinate collective action. By creating one platform together, it helped stakeholders with different, and sometimes divergent, interests to find collective win-win agreements, rather than creating inappropriate private platforms as they may have in the past.

Regarding the design of the wine trail signage, new road legislation in France that required harmonization of the shape and color of signs, (which at first seemed to be an administrative constraint) ended up helping all parties to come to agreement much faster by reducing options. This meant that collective coordinated action had far more chance of successfully placing signboards than separate initiatives.

The Brand: Provence

One of the most important outcomes was to add value to the brand and make the destination of Provence more attractive. The identity, culture, and values of Provence wine led to a stakeholder consensus around the Provence brand for both wine and tourism. Provence and the wine brands thus created synergies for the promotion of wine tourism, as well as a bridge between the two worlds.

The importance of this level of consensus around regional brand is highlighted in the wine tourism literature. Experts attest that both the wine and tourism industries rely on regional branding for market leverage and promotion (Fuller 1997; Hall and Macionis 1998; Hall et al. 1998). Tourism is fundamentally about the differences between places (Relph 1996), while wine is "one of those rare commodities which is branded on the basis of its geographical origin" (Merret and Whitwell 1994: 174). According to Hall (1996): "There is a direct impact on tourism in the identification of wine regions because of the inter-relationships that may exist in the overlap of wine and destination region promotion and the accompanying set of economic and social linkages (p. 114)."

Agreement on Geographical Borders

Another best practice, and unexpected benefit of the Provence Wine Trail, was the inclusion of the "French Riviera" players, who were originally going to be excluded from the project. Because this part of Provence didn't have a tourism information center, and was governed by an organization of wine producers and dealers, it was possible for the CIVP to operate as a partner. In other words, because of its legal status, the CIVP mission was able to overcome any administrative barriers.

In this way the CIVP is similar to a tourism product club, defined as "*a group of companies that have agreed to work together to develop new tourism products or increase the value of existing products and collectively review the existing problems that hinder profitable development of tourism*" (Del Campo et al., 2010). The three main lines of development of the tourism product club are: (1) Communication about the tourism enterprise, (2) Education for the tourism industry, (3) Research into segment needs. Thus, the Provence Wine Trail includes wineries from the Alpes-Maritimes to promote wine tourism in the south east of France. The stakeholders agreed to combine their assets and allow the CIVP to promote all of Provence, rather than to rely only on regional organizations. This was a new approach in the French tourism landscape.

FUTURE ISSUES

The Provence Wine Trail still has to tackle new challenges with regards to the organization of wine tourism and changing trends in tourist behavior. One of these issues is additional training sessions for winemakers, such as language courses, tourist customer service courses, regulations on safety, security, and accessibility, tastings, and interior decoration.

Obviously continued focus on marketing and promotion is always needed. Stepping up the Wine Trail promotion with more press conferences and tastings for international trade, as well as translating brochures and leaflets into English are all important actions. In addition, encouraging stakeholders to update the database so the website stays timely is part of this, as well as developing partnerships with new tour operators.

Continued focus on wine tourist behavior and new trends is also critical. For example, some of the new tourism trends include:

- **Shorter Vacation Breaks**: Fewer and fewer people have enough free time to take long vacations. This means there is an opportunity to organize more frequent and shorter trips. Consequently, winery owners should be prepared to receive tourists year round.
- **Health**: Health is a major concern for tourists. The development of spa and sports activities is a new challenge for wineries and vineyard owners.
- **Environmental issues**: Tourists are increasingly concerned with the environment. For instance, wine growers offering biodynamic and natural wines provide a better match with tourist expectations.

DISCUSSION QUESTIONS

1. Identify the major challenges a new wine region has when developing a wine trail.
2. Create a list of all stakeholders who need to be involved in development of a wine tourism regional brand.
3. Conduct a SWOT analysis on the Provence Wine Trail as it is today.
4. What other actions does the Provence Wine Trail need to take in order to continue to attract tourists to the region for the next decade?

Chapter Eight

Città del Vino: A National Effort to Promote Wine Tourism in Italy

Roberta Capitello, Lara Agnoli, Ilenia Confente, Paolo Benvenuti and Iole Piscolla
University of Verona and Associazione Nazionale Città del Vino, Italy

October 24, 2014 was an important date for Città del Vino, which stands for Associazione Nazionale Città del Vino, or National Association of Wine Cities in Italy. It was the date that Paolo Benvenuti, Director of Città del Vino, and Tian Xiangli, representing the city of Qinhuangdao in China, signed an agreement. They were very excited during the signing because they were aware that a new stage in the history of Città del Vino was being formalized to encourage exchanges and international projects with foreign cities in matching tourists with wine regions.

Another important day was March 21, 1987, when 39 mayors met in Siena to establish the Città del Vino. The founders represented the rich mosaic that constitutes the Italian vineyards, including small and large municipalities from the north to the south of Italy, some of them already known in the wine world, while others were still in the shadows.

This was a very important occasion, because the Italian wine industry was suffering greatly as a result of the methanol contamination scandal in 1986. This negative event represented one of the main reasons to establish the Città del Vino. The founders understood that the rebirth of the wine industry was inextricably linked to a cultural renaissance that would strengthen the link between wine and territory that characterizes the originality of Italian wine.

Since then, the goal of the Città del Vino has been to assist municipalities in developing activities and projects concerning wine, food, and local products that better the quality of life and promote sustainable development, tourism and work opportunities. However, working together as a unified entity was not always easy. This chapter explains how Città del Vino came together as a new concept to help unify the wine industry in Italy and promote positive wine tourism practices. It describes the collaborative actions of multiple stakeholder groups to make this concept successful.

OVERVIEW OF THE WINE INDUSTRY IN ITALY

According to the OIV (2015), Italy produced over 44 million hectaliters of wine in 2014, making it second in production to France. Italy boasts 645,000 hectares of vineyards (OIV, 2015), and approximately 160,000 wine producers (Unioncamere, 2009). Also considered to be one of the largest wine exporters in the world, Italy shipped nearly half of their total production of 20 million hectolitres with a value of $5 million euros in 2014 (OIV, 2015) to many other countries around the globe.

# of Wineries in Italy	160,000

Italy is home to a large number of unique varietals, but its signature grape is sangiovese. The graph illustrates that sangiovese also has the largest production, followed by montepulciano, merlot, cataratto commune, trebbiano toscano, barbera, chardonnay, prosecco, pinot griogio, and calabrese (Istat, 2011).

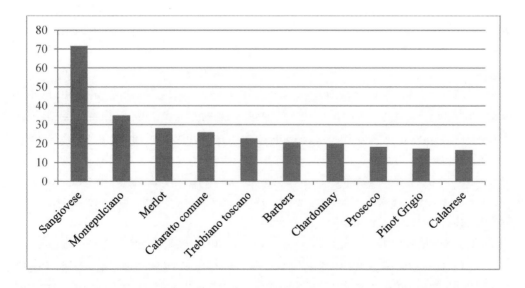

Top 10 Italian Grape Varietals (thousands of hectares of vineyards)

Italy is characterized by a high number of Protected Designation of Origin (PDO) wines that represent about one third of the total production. In 2014, there were 405, with 73 designated DOCG and 332 designated DOC (Federdoc, 2014). Piedmont, Tuscany and Veneto feature a large number of PDOs and significant production, while other regions have a smaller number of PDOs, but are focused on the supply of traditional products. In Italy, 118 PGI (Protected Geographic Indication) wines were produced, consisting of slightly less than 30% of the entire wine production.

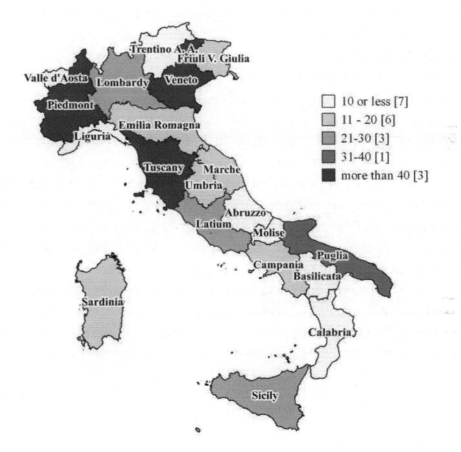

Map of Italy Showing PDO Wines by Region

Wine and Tourism in Italy

Italy is one of the most visited countries in the world (ONT 2014). This success has been mainly linked to natural, architectural, and artistic attractions, but the focus in terms of tourism has been primarily on a few coastal and urban destinations. Many rural spaces and small urban centres have been completely left out of the tourism flows, although they have high potential to attract tourists thanks to their rich cultural heritage, relaxing atmospheres, landscape features, and regional food and wine specialties.

However, this is evolving due to the changes affecting the tourism industry in Italy. According to the Natsional Tourism Observatory (ONT 2014), the economic crisis has drastically reduced domestic demand, which was dominant in terms of consumption, added value, and employment. Indeed domestic tourism decreased by 40.9%, in terms of overnight stays, from 2005 to 2013 (ONT 2014).

Despite the domestic drop in tourism, Italy still maintains its importance at the international tourism level, remaining in fifth position, following France, the United States, China, and Spain, with an increase of 27.1% from 2005 to 2012 (ONT 2014). In particular, the demand for short holidays and the preference for new types of tourism, from health and sustainable tourism to religious tourism, or rural and gastronomic holidays have increased (UNWTO 2014). Wine tourism is an important source of value for small firms and local communities in Italy. According to the estimations of the National Tourism Observatory (ONT, 2012), food and wine tourism generated 5–6% of arrivals annually during the period 2008–2011.

Despite its history and cultural heritage, Italian tourism suffers from difficulties in maintaining regional resources, in offering a unique identity for tourism initiatives, and in integrating different destinations. Moreover, the link between wine and destination brands has not yet been taken into account by the Italian Destination Marketing Organisations (DMOs). Wine tourism is further penalized by the fragmentation of the industry and a predominantly individualistic vision of wineries and wine regions. This has led to low impacts for individual initiatives, a high risk of failure, and many investments with low efficiency (Cavicchi & Santini 2014). The Italian wine industry has to deal with this crisis and the lack of resources devoted to cellar door marketing.

Nevertheless, over the last decade, wine tourism has increased in Italy thanks to the enhancement of hospitality at wineries, the flourishing of cellar door initiatives (such as wine tasting, theatre or music in wineries and wine museums), and the improvement of some critical success factors through collective initiatives. Furthermore, many diversified festivals and events reflect local traditions in almost all of the Designations of Origin areas. These have followed the increasing attention placed on communication by the Consortia of Protected Designation of Origin wines and the development of new 'Strade del vino e dei sapori,' or Wine and Food Roads, of which there are now approximately 160.

The Città del Vino: Purpose and Objectives

The Città del Vino is a non-profit association of local institutions, based in Siena. It was created to defend and develop the quality of production and the territories in the cities and villages that have a wine industry from economic, social and cultural viewpoints. To this end, it implements actions, initiatives and services in the protection, promotion, and information fields. Today, the Association has 469 members; mainly municipalities (441), but also unions of municipalities, provinces, wine roads, and natural parks and associations.

Map of the Cities of Wine in Italy, 2014

The Cities of Wine represent about 5% of the 8,092 Italian municipalities. They are located throughout Italy, with particularly high concentrations in traditional production areas. They are often small municipalities, where winemaking is the center of economic and social life. Examples include Barolo, Valdobbiadene, and Montalcino, as well as areas known for hospitality and tourism such as Rimini, Ravello, and Sirmione.

Major towns that are active in wine tourism, such as Grosseto, Trento, Bolzano, and Gorizia, are also included, as well as major cities that recognize food and wine as an increasingly important component of the tourist experience, such as Rome, Siena, Palermo, and Perugia. Altogether the associated municipalities represent 70% of Italian vineyards, about 45% of PDO vineyards, and 12% of the tourism accommodation supply.

The purpose of the association is to strengthen the link between wine and territory, as stated by the preamble of the Statute:

> *"The protection of the quality of products (wine and typical products) and the quality of territories (environment, landscape, quality of life), their development and their promotion are essential prerequisites for any possible policy of growth and development; therefore, they must be the focal points in the actions of public administration and businesses; Città del Vino, as a network of wine territories, is a fundamental part of this idea of development, with the active participation of member municipalities, economic and social institutions, farms and wineries, crafts, commerce, culture, and hospitality. Through Città del Vino, this network can consolidate its role of political and institutional intermediary at local, regional and national levels and become a participation and planning instrument to promote the territorial development, sustainability, and solidarity."*

The main objectives of the Città del Vino are listed in the Article 3 of the Statute and include:

1) To improve and protect viticultural quality, local and ancient vines, traditional architecture and farming practices, landscape and local products, avoiding the use of GMOs;
2) To raise awareness about and directly involve people in the care, protection, and improvement of the territory as a common, cultural and identity heritage;
3) To promote responsible and moderate drinking and proper nutrition, as well as consumption ethics.

From an organizational point of view, Città del Vino is structured in a general meeting of members, a national council, which elects the president and two vice presidents, and regional and interregional coordinating committees. The Executive Office coordinates all activities. Città del Vino is mainly funded by the associated municipalities, which are required to pay an annual membership fee.

THE PROBLEM: LACK OF COLLABORATION AND COMMITMENT FOR REGIONAL WINE TOURISM

Since its establishment, the Città del Vino has been aware of many limitations affecting wine tourism development in Italy. The main issues stem from structural characteristics of the Italian wine industry, namely micro- and small-sized firms with an individualistic approach to marketing and little propensity to collaborate.

Another issue is the lack of attention to wine tourism hospitality. In the past, wine sales were primarily bulk wine sold to local and traditional consumers. Commercial development towards wine bottling took place through the long supply chain by intermediaries, to reach both the entire national market and an increasing number of foreign markets. Therefore, cellar door hospitality to wine tourists was neglected.

The small size of many Italian wineries also contributed to a lack of hospitality culture, because they didn't have the staff or training to provide positive cellar door experiences and sales. So even though local wine associations developed promotions to bring visitors to the region, there was often little support in developing hospitality quality and cellar door attractiveness.

Finally, there was little collaboration between Italian wineries and other local food producers or businesses promoting historical, artistic, and cultural heritage, or rural and environmental attractions. This opportunity to work together and support one another was missing, yet it could be the pivotal component in preserving and developing the rural regions to increase tourism attractiveness.

Therefore, the challenge for the Città del Vino was how to overcome these problems and get everyone on board to support Italian wine tourism. Municipalities were especially challenged by the need to get all players in the wine and tourism industry to agree on the following issues:
- Enhance territorial identity in the local community
- Improve the tourist hospitality experience

- Increase the quality of life for residents
- Achieve business efficiency and economics using a rural tourism model
- Integrate the various tourism industries to support one another

THE SOLUTION: CITTÀ DEL VINO IMPLEMENTS AN ACTION PLAN TO GAIN COMMITMENT FOR TOURISM

In order to address the many issues before them, the planning committee for Città del Vino created a unified plan that inspired commitment from all parties. This included the development of a charter and a regulatory plan, establishing a partnership with Movimento del Turismo del Vino, creating special events and activities for wine tourism, and implementing unique initiatives such as Tuscany Wine Architecture and an international wine competition. The following section describes how they executed this solution.

Step One: Establish a Charter

The 'Charter of Quality' was the first step Città del Vino took to deal with the main Italian wine tourism challenges. The Charter was intended to introduce the principle of promoting member municipalities based on their efforts and activities to improve their recognizability as a City of Wine. According to their specific characteristics, history and traditions, the Cities of Wine follow the same set of requirements in order to reinforce their identity and distinctiveness compared to other cities. The following table outlines the ten principles of the Charter of Quality that inform the institutional decisions of each associated municipality.

The Ten Principles of the Charter of Quality for Città del Vino

1. To protect the wine landscape	The rural landscape has taken form through the centuries and has its own characteristics, identity, and harmony. The City of Wine should protect the landscape and the vineyards; respect the geomorphology of the territory including support and development that does not harm the landscape or the vineyards.

2. To simplify the administrative procedures for the wine industry	An objective of the City of Wine is to simplify the administrative procedures necessary for the adaptation of production and agritourism structures or for business and housing expansion, to ensure that they are carried out in a timely manner, according to business needs.
3. To enhance wine perception	Visitors need to recognize that they are in a wine producing area at first sight and that they are on a 'wine road', where they can find not only vineyards and wineries, but also wine tasting, wine shops, museums or exhibitions, restaurants, farm hospitality, and other relevant activities.
4. To promote wine culture	The wine culture is also expressed through the enhancement of wine memory. The City of Wine aims to promote museums for wine and local rural culture, private and public related to wine production, and oral or written sources linked to wine business and consumption.
5. To invest in the wine roads	The wine road is the 'glue' linking regional winegrowers, wineries, and other businesses. It is a factor of the distinctiveness of a wine region. The City of Wine should enhance and promote its 'wine road' to give more value and visibility to the territory and local partnerships.
6. To set up a territorial 'Enoteca' (a local wine shop)	Visitors should be able to find one or more public and private locations where all the local wine production is shown. This is to let visitors learn about the wines, producers, and their characteristics.
7. To promote local wines in restaurants	Local wines must have a prominent place in the wine lists of local restaurants. The City of Wine should promote the inclusion of territorial and traditional dishes on menus and their matching with local wines.
8. To promote sustainability in the wine industry	Wine production in the City of Wine should follow environmental and health-friendly principles and techniques. The City of Wine should promote sustainability by producers during the entire production cycle.
9. To promote creativity around wine	Wine has inspired artists over the centuries. The City of Wine should steadily solicit artistic and cultural expression around wine through prizes, competitions, festivals, exhibitions, events, and initiatives.
10. To set up a wine calendar	Winegrowing follows its own calendar marked by the work phases in the vineyard and the cellar. The City of Wine should set up an annual calendar of events to raise awareness of its wine and food production.

Step Two: Develop a Regulatory Plan

The Regulatory Plan encourages municipalities towards best practice in urban and rural planning by following a set of environmental and socio-economic requirements. The main principle of the Plan is that socio-economic development, such as housing, infrastructure, and production areas, needs to be designed to integrate with the landscape and its relationship with winegrowing. Preserving the rural environment is a primary objective.

Città del Vino provides local administrators, producers, and urban planners with methodological guidelines to promote sustainable governance of a territory. The goal is to recognize the value of the vineyard system and its fragility, and to only develop projects that are committed to sustainability. Over time, Città del Vino has continued to update strategies and implement initiatives to recognize the commitment of the best performing municipalities and disseminate best practices. Following is a list outlining elements, tools, and initiatives:

Basic Elements of the Regulatory Plan for Cities of Wine

Basic principle: The countryside expresses values, as the city does. The needs of the countryside are not subordinate to those of urban areas.
Primary tool: To extend knowledge in order to: • Identify the most suitable wine territory;   • Protect them from unsuitable locations;   • Identify the vulnerabilities of soils and develop rules to ensure the sustainable use of agricultural land and water control;   • Develop good practice together with the wineries;   • Explore ways to reduce conflict between alternative uses.
Pivotal Element: The relationship between population and producers. Listening to and actively involving residents is crucial because they are the keepers of knowledge and the authors/actors of the territory and the landscape.
Supporting Initiatives: • Città del Vino enriched the Regulatory Plan with specific content in terms of landscape, measures to deal with climatic deterioration, updating of cultivation techniques for vineyards, quality of rural architecture, and their reflections on the territorial government.   • The Landscape Project provides municipalities with operational guidance to integrate the landscape in municipal instruments.

- The document *'Wine and Landscape: Materials for the Territorial Government'* was published in 2009 to support local administrators, producers and urban planners.
- The document *'Resolving the Conflict between Agricultural Use and Energy Use of the Agricultural Soil'* was published in 2011 to provide local administrators with directions on renewable energy sources.

Every two years the Competition for the Best Regulatory Plan of the Cities of Wine is held in collaboration with the National Institute of Urban Planning. It is aimed at municipalities and territorial bodies using an instrument of urban and regional planning attentive to the sustainable development of the territory and the protection of wine areas.

Step Three: Establish a Partnership with MTV Calici di Stelle

Città del Vino decided to create a collaborative partnership with Movimento del Turismo del Vino (MTV) to participate in the event Calici di Stelle, which stands for "Night of the Shooting Stars." Calici di Stelle is one of the most anticipated summer events by wine tourists and wine lovers. Every year, for several nights in August, wine is celebrated throughout Italy. With the two organizations working in partnership, the event now brings to life the historic centers of the Cities of Wine, and well as all of the local wineries.

MTV supports the municipalities and the wine roads in the organization of the event by providing them with services, such as advertising, communication, supplies, organizational services, and guidelines to be followed by all participating municipalities. They offer a common image and bring high standards to the event, which occurs throughout Italy.

Step Four: Create Special Activities in Wine Cities

To support the Calici di Stelle event each year, the Cities of Wine create a variety of activities focused on wine. These include wine tastings in municipal squares with the event logo on all glasses, as well as many activities to present the traditional wines and unique varietals. There are Italian music concerts, open museums, and many other types of entertainment designed to introduce the public to the local culture, the cellar door and the territory. Some cities offer guided observation of the stars through telescopes with the sponsorship of the Italian Amateur Astronomers Union. Care is taken to promote responsible consumption of alcoholic beverages in all activities.

Flyer Cover for Calici di Stelle.

Another special activity that is part of this event is the photography award '*Federica's Star: Best Picture of the Year*'. It is designed to enhance the relationship between the event and the surrounding environment, and provides recognition for the city in which the winning picture is taken. Following are the winning photographs for the 2014 edition, which also show the festive atmosphere during the event in the Cities of Wine.

1ˢᵗ Place Photo by Nicola Sesto (Lamezia Terme, Calabria)
*'For the ability to combine the evocative power of the festival with the hosting
site of Calici di Stelle in a skillful interlacement of tales and dreams.'*

***2ⁿᵈ Place Photo by Strada del Vino Nobile di Montepulciano
(Montepulciano, Tuscany)***
*'For the ability to skilfully combine the municipal square element, the vital
heart of the Cities of Wine, with the event Calici di Stelle.'*

3rd Place Photo by Gian Carlo Casula (Verucchio, Emilia Romagna)

'For the ability to capture the spirit and environment of the summer festival promoted by the Cities of Wine: the municipal square and the young generation.'

Step Five: Implement Unique Initiatives

Città del Vino also undertook implementing several initiatives designed to promote wine tourism in Italy. These centered around architecture and gaining more international recognition.

Highlighting Tuscany Wine Architecture - A major initiative implemented by the Città del Vino was the introduction of *Tuscany Wine Architecture*. This concept highlights a network of 25 wineries that were designed by well-known Italian and foreign architects or enriched by artworks of the twentieth century. They are mainly concentrated in Tuscany, in the provinces of Livorno, Grosseto, Arezzo, Siena, Florence and Pisa, forming a contemporary art and design route that is unique in Italy.

These wineries are buildings of high architectural quality designed over the last 25 years using innovative construction and production technologies. They possess a renewed aesthetic expression, design choices promoting green architecture, innovative integration between technologies in the field of energy, reduction of the visual impact, and a perfect integration with the landscape. The presence of modern artworks in these wineries has developed a new arts patronage for contemporary artists and generated an innovative system of culture and production. They represent a real treasure for the artistic and cultural heritage of their territory, and have opened their doors for tours and tastings.

Establishment of International Partnerships - More recently Città del Vino has expanded outside of the borders of Italy to establish partnerships with international wine cities. It became one of the promoters of the European Network of Cities of Wine, RECEVIN, established in Strasbourg in 1999. It aims to enhance trade relations between European wine producing areas and to strengthen the voice of the Cities of Wine. It includes cities of wine from Portugal, Spain, France, Austria, Italy, Slovenia, Germany, Hungary, and Greece, and organizes annual internships for young winemakers.

Creation of an International Wine Competition - Another creation of the Città del Vino is the *Mayor's Selection*, which is an international wine competition to promote wineries and territories that produce PDO or PGI quality wines in small quantities (1,000–50,000 bottles). The unique aspect of this initiative is that the winery and the municipality where the vineyards are located take part in the competition together.

The Mayor's Selection is a traveling wine competition, in collaboration with RECEVIN, the European Network of Cities of Wine. It changes the tasting venue each time the event takes place, in order to promote all the territories of the Cities of Wine. The competition is linked to other initiatives supported by the Città del Vino to promote specific wine typologies such as sparkling wines, raisin wines, wine fermented in clay pots, organic wines, or wine made using sustainable farming practices. The 2015 edition took place in Lisbon, Portugal and involved more than 1,100 wines, with about 700 from Italy.

RESULTS AND WHY CITTÀ DEL VINO IS CONSIDERED A BEST PRACTICE IN WINE TOURISM

The experience of the Città del Vino can be considered a best practice in wine tourism for several reasons.

Collaboration with Regional Municipalities – Città del Vino identified the rural municipality as the most important actor to promote wine tourism, even though it is the smallest public decision maker. Because the regional municipality is the closest public institution to the wine region, it can ensure the highest knowledge of the characteristics of the economic, social and cultural systems within this region. In this way, it can ensure the most appropriate measures are taken to promote the protection of the natural and cultural heritage, as well as to support the development of a wine hospitality culture. The municipality's commitment to the Città del Vino illustrates that they recognize the focal role of wine in promoting rural development and tourism.

Creation of a Nationwide Network – Another reason is the ability of the Città del Vino to create a nationwide network of municipalities. This allows it to play an important regulatory role by proposing guidelines and benchmarks. Città del Vino provides support to municipalities in their territorial planning and in rural environment preservation. This allows each wine city to develop a unique identity and to elevate the quality of wine tourism.

Establishing Uniform Standards for Wine Hospitality – By providing guidance and support to municipalities in establishing wine tourism, Città del Vino has been able to ensure uniform standards in wine region management and hospitality.

Expanding Relationships and Partnerships - Città del Vino has also worked to establish relationships outside the network of municipalities by involving public and private organizations that directly and indirectly participate in wine tourism. Examples include partnerships with MTV Calici di Stelle, the Tuscany Wine Architecture project, the European Network of Cities of Wine, (RECEVIN), the Mayor's Selection, professional conferences, research collaborations, and many projects with wine and food roads.

Economic Return - The success of these initiatives has helped Città del Vino to not only promote economic return on investment, but also to strengthen the attractiveness and competitiveness of the wine cities. By providing local institutions with services, communication, promotion activities, and information, Città del Vino helps the success of wineries' investments in wine tourism.

Achieving Broader Recognition in Italy - These abilities have enhanced the power of the Città del Vino in its relationships with public institutions, such as the Ministries of Agriculture, Environment and Tourism, the Agriculture Committees of the Italian Chamber and Senate, regional administrations, Chambers of Commerce and local action groups.

In summary, Città del Vino's activities represent examples of good practices for other organizations and wine regions aiming to generate new wine tourism opportunities. By adopting these practices other wine regions can strengthen the link between wine businesses and the local communication, and also preserve local traditions, the rural landscape, and promote a hospitality culture with high quality standards.

FUTURE ISSUES

Città del Vino has been very successful in promoting wine tourism in Italy, but still faces certain challenges. A major one is to continue to increase the quality of wine tourism and meet tourists' expectations of the wine and food hospitality industry in Italy. This requires a continued focus on collaboration over individualism, and reinforcing the role of the municipalities in tourism planning and implementation.

Other important issues include the ability to continue to develop best practices and support their implementation across regions, while at the same time respecting different city needs. For example, the Tuscany Architecture program works well in Tuscany, but other cities may have different areas of focus and pride. It is for this reason that Città del Vino recognizes the importance of giving voice to the needs of territories and local communities, because they are the most important defenders of local typicality, cultural, food and wine, environmental, and landscape traditions. All of these factors constitute added value for Italy.

DISCUSSION QUESTIONS

1. What values does Città del Vino wish to pursue and disseminate through its work?
2. Perform a SWOT analysis on Città del Vino.
3. What role does Città del Vino give to the municipalities in the promotion of wine tourism in Italy, and should this be enhanced?
4. What advantages can events like Calici di Stelle offer for the development of wine tourism in Italy? What risks can be seen?

5. What new actions and initiatives would you suggest to Città del Vino to strengthen its ability to network and communicate with the market and public institutions?

Chapter Nine

Wine Tourism on an Isolated Island: The Intriguing Case of Waiheke, New Zealand

Nick Lewis & Lucy Baragwanath, *University of Auckland*

A perfect Sunday in Auckland, New Zealand – hot sun, sparkling harbor, and the islands of the Hauraki Gulf in the distance. Crowds gather at the ferry terminal, excited at the prospect of a thirty-five minute harbor cruise to Waiheke Island. Getting off the ferry to the hustle and bustle of Matiatia Wharf, everyone is now on island time. A fabulous day of food, wine, and sun lies ahead. A wine tourism manager describes the allure of Waiheke Island:

> *"What is the allure? It's a special place. A destination. A beautiful island, great wine, a well-known winery destination. There's the critical mass, many vineyards, and the quality is good."*

The picture conceals two, closely connected, compositional flaws – the day's experience can be interrupted by the challenge of getting around, and the sun doesn't always shine. According to a survey of wine tourists visiting Waiheke Island in the summer of 2009, some of the frustrations identified were getting to and from cellar doors and being unable to visit multiple wineries (Baragwanath & Lewis, 2009). The manager of a winery on Waiheke Island sums up some of these issues:

> *"They get off the wharf, and what do they do then ... there're no signs, the buses don't go to the vineyards ... they can be left confused and stranded."*

In winter, when the ferry ride can lose its charm, the vines are bare, and rain is imminent, disembarking from the passenger ferry at Matiatia Wharf without a plan for getting around can come as more of a shock. Seasonality is an endemic problem in wine tourism, and tends to present different problems in different settings (Hall 2005). On Waiheke, it intensifies the paradox of '*proximate isolation*'.

Though only 17 kilometers from the major metropolis of Auckland, the island can sometimes feel isolated for tourists coming to visit the wineries, restaurants, shops, and sites. This chapter explores how Waiheke Island has moved forward in solving some of the logistical and seasonal issues that hinder wine tourists from enjoying their time on the island.

OVERVIEW OF NEW ZEALAND WINE

Wine grapes were first planted on the North Island of New Zealand in 1836 by James Busby, a British colonist (Robinson, 2006). Today there are 698 registered wineries and 833 growers in New Zealand (NZ Wine, 2013), as well as a complex of intermediaries such as distributors, wine writers, specialist retailers, brokers, investment funds, wine and viticulture teaching, and research programs. The wineries and vineyards lie at the heart of the industry, and are made up of a group of globally owned corporate operations, mid-sized often domestically owned and operated wineries, and a large number of small, often family based operations. Similarly, the vineyards range from small retirement investments to major viticultural operations.

New Zealand has 10 main wine regions, the most renowned of which are Marlborough and Central Otago, which are famous for the production of sauvignon blanc and pinot noir wines. Dominated by cool climate wines, these regions stretch from latitude 35 to 43 S and embrace a wide range of climatic conditions from continental summer heats and cold nights to more maritime and relatively moist conditions. While definably cool climate in many places, the Hawkes Bay, Waiheke, and Northland regions are now making syrah and other warmer climate wines, often to critical acclaim.

By international standards the New Zealand wine industry is small and export-focused. In all, New Zealand exports 169.6 million liters of wine annually, of its total production of 248.4 million liters, at a value of $1.211 billion (NZ Winegrowers Annual Report 2014). Today its most important markets by volume are Australia, United Kingdom, and the United States with Canada a distant fourth. China is a growing market and a significant target for New Zealand producers.

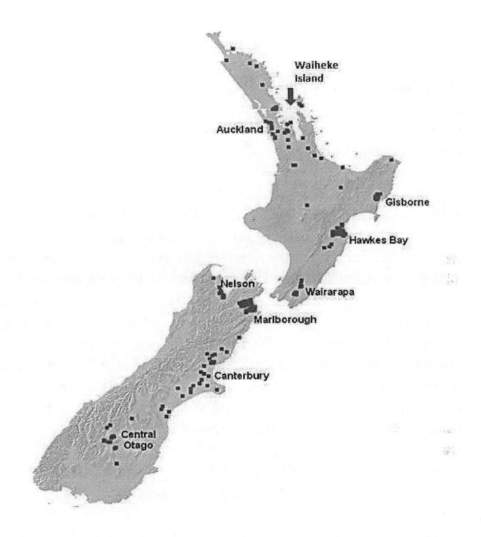

Map of New Zealand Wine Regions

For the last 25 years New Zealand's wine industry has also been characterized by spectacular growth. It went from producing 56.4 million liters in 1995 to 248.4 million liters by 2013 (Hayward and Lewis, 2012, Lewis 2014). This growth was built on two pillars: high volume production of Marlborough Sauvignon Blanc, largely for export; and increased production of a variety of super-premium wines for domestic and international markets from across New Zealand's wine regions. Alongside these two pillars, New Zealand has led the production and export of affordable premium and super-premium Pinot Noirs, again largely from Marlborough, which now dominates the industry's production by volume. The industry has now stabilized and matured into a major national export earner known for high quality wines and is a significant, albeit relatively small, participant in the world of wine.

Wine Tourism in New Zealand

Tourism is a major industry in New Zealand, and is centered on its landscapes and outdoor activities. NZ has become a major international destination for this form of tourism, and wine has become an increasingly important part of that proposition. Wine tourism figures are not routinely collected, but estimated numbers of wine tourists peaked at just over 600,000 in 2007, with international tourists making up a third. International wine tourists spent approximately $3500 per person in 2008 in New Zealand, amounting to $700 million (Ministry of Tourism, 2009).

Central Otago in the South Island and Hawkes Bay in the North Island are the most celebrated wine tourism regions and support an extensive wine tourism infrastructure. Central Otago is a year-round international tourism destination, and at the heart of New Zealand's iconic alpine landscapes known for its lakes and mountains. Hawke's Bay is a prominent summer tourism destination, particularly for domestic tourists and its wineries and vineyards are a pivotal part of the attraction.

# of Wineries in New Zealand	698
# of Wineries on Waiheke Island	30

Auckland, however, is the most visited wine region. It is gateway to New Zealand for international visitors and by far NZ's largest city with 35% of the national population, yet offers few daytime attractions. Its wineries are an important part of its domestic and international tourism, and are widely spread, but cluster into five distinctive sub-regions: 1) Matakana in the north, 2) the West Auckland regions of Huapai, and 3) Kumeu, where much of the New Zealand wine industry first developed, 4) Clevedon in the south, and 5) Waiheke Island.

Wine tourism is especially crucial to many of the smaller wineries in New Zealand. In 2014 cellar door sales made up 7% of total revenue among those wineries with revenues between $1.5 million and $5 million per annum (Deloites 2014). For wineries such as those on Waiheke, where a large proportion of this revenue stream accrues as profit margin, this represents a significant dollop of cream on top of wine sold through normal commercial channels and is often the income that allows them to remain profitable.

OVERVIEW OF WAIHEKE ISLAND

Thirty years ago Waiheke Island was a sleepy and rural holiday destination for Aucklanders and a home to retirees, alternative lifestylers, and a small fishing and agricultural community (Picard 2005; Monin 1992; Lonely Planet 2012). While only 17 kilometers from Auckland in terms of distance, it was a long way away culturally. In the intervening years, while some of these features remain, it has been transformed into an upmarket commuter suburb where many professionals now work from home, and a tourism destination centered in significant part on its wineries. Wine has become a prominent feature of the economic and physical landscapes and one in three tourists to the island visit at least one vineyard during their day out to Waiheke (Baragwanath and Lewis, 2009).

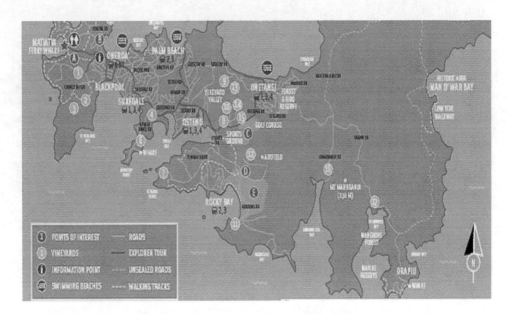

Map of Waiheke Island Wineries & Places of Interest

Today there are around 30 wineries across the island, although clustered in the west. Most are very small with an average planted area of five hectares (WWA, 2006), and produce less than 1,000 cases per year. Man o' War Estate is the only large producer with 120 acres of grapevines planted on a 4500 acre station that takes up a large part of the largely unoccupied eastern end of the island.

Syrah and Cabernet blends are the region's flagship wines, but wineries produce wines across the range of New Zealand styles. Wineries such as Jurassic Ridge and Obsidian experiment with varieties such as Montepulciano (Kelly 2009). Several produce wines from grapes grown under contract in other regions. For almost all, operating profitably under low-yield, low-volume, high cost conditions is a struggle. Wine tourism is very important, as it can comprise in excess of 50% of revenue for some enterprises, while a number report that up to 20% of their wine sales are generated through the cellar door (Baragwanath and Lewis, 2014).

Cable Bay Winery Vineyards on Waiheke Island

The first commercial vineyard was planted in 1978 by Kim and Jeanette Goldwater, with Stonyridge establishing its first vineyard in 1982. By the mid-80s, Goldwater and Stonyridge were producing highly regarded Bordeaux-style wines on the islands. Others followed in the 1990s and early 2000s, attracted by the island's lifestyle and winemaking opportunities and by the relatively low land values for these features so close to Auckland. The arrival of fast ferries in 1987 suddenly made the island accessible for daily visiting and commuting (Baragwanath, Lewis and Priestley 2009). Today the cost of land has risen considerably.

The Winery Scene on Waiheke Island

Four general types of wine enterprise have been identified on Waiheke (Baragwanath and Lewis (2014):

1. Passionate owner-operator family winemaker;

2. Hospitality enterprise for which vineyard and winery are crucial, but often not the primary feature of the business as a whole;

3. Wine export-focused high-end wine enterprise making distinctive super-premium or iconic Waiheke wines;

4. Mid-sized winery producing premium and super-premium wines for domestic and export markets.

While each type of enterprise has its exemplar (Jurassic Ridge, Mudbrick, Destiny Bay, and Man O'War Estate, respectively), in reality different enterprises embody dimensions of several of the types. The iconic Stonyridge, established by internationally renowned yachtsman and adventurer Stephen White, made its reputation from its award winning wine Larose as well as White's own profile, the Stonyridge vineyard café and summertime dance parties. Today the winery is a serious winemaker, a hospitality company, and a mid-sized producer that makes and serves at its café and cellar door a second label 'Fallen Angel by Stonyridge' made under direction of its winemaker but sourced from and produced in other regions. In contrast, Te Whau serves its own high-end wine and offers sophisticated dining experiences in an architecturally designed restaurant with stunning views. Cable Bay does the same but produces higher volumes of wine for domestic and export markets, much made from grapes sourced in other regions.

Waiheke wine enterprises suffer the business challenges faced by small family enterprises everywhere (Deloitte and New Zealand Winegrowers, 2014). These are intensified by high cost, low yield production conditions, and the challenges of distributing super-premium priced wines from the relative isolation of the island in a small and crowded domestic wine market. Nonetheless, Waiheke wine has proven a successful proposition. The 'region' has a defining reputation for higher quality, higher priced wines, and distinctive physical and cultural environments. Centered on its islandness and profiled through its tourism, this distinctiveness allows for rich provenance narratives that emphasize terroir, cultural history, and exclusivity. As the Island of Wine website observes:

> *"Sunshine, sea breezes, and rolling hills ... define one of the world's most exclusive wine growing regions, producing wines with intense varietal flavor and the freshness and purity that comes from a pristine environment. There is nowhere in the world like Waiheke Island. Discover its beauty by discovering its wines...* (waihekewine.co.nz; 2014)."

All Waiheke wineries draw on wine tourism, whether directly through the cellar door or hospitality initiatives or indirectly through its marketing and reputation enhancing effects. Even those wineries that do not open a public cellar door benefit from the experience and memories of Waiheke generated from tourism, the media coverage and profile that it generates, and sell their wines to high-end tourists through retailers and restaurants. Wines are relatively highly priced and while most wines sustain those prices with a quality reputation, the added appeal of purchasing wine from a cellar door or consuming it on-site at a vineyard restaurant is crucial to enterprise strategies.

THE PROBLEM: ISOLATION DUE TO SEASONALITY AND TRANSPORTATION ISSUES

Winery managers and tourism operators on Waiheke Island knew how critical tourism was to their success, but they were plagued by the two major issues of seasonality and transportation.

Seasonality - In terms of seasonality, keeping a flow of tourists through the cellar door was critical, but Waiheke suffered from extreme seasonality. The vast majority of tourists visited between November and April, which is spring and summer in the southern hemisphere of New Zealand. During the other months of the year, visitor numbers were very low. This had particular problems for sustaining menu and service quality in the high-end winery restaurants where the financial costs of staying open had to be balanced against the reputational costs of opening with skeleton staffs or closing seasonally and losing key staff. Research shows that keeping a cellar door open is costly in terms of facilities and staffing, and it can divert energy and resources away from other activities (Hall, 2004; Baragwanath 2009).

Transportation – There was also a lack of independent transportation to take visitors to the wineries. This issue was amplified by the dispersed location of wineries on Waiheke. It was exacerbated further by the routine closure of cellar doors on particular days or in winter months. Furthermore, growing competition from other wine regions closer to Auckland, such as Matakana that didn't require a ferry ride, were beginning to intensify their problems.

While Waiheke wineries and tourism enterprises were searching for answers, researchers from the University of Auckland conducted a tourism survey in the summer of 2009. The researchers interviewed 40 people from different agencies, industries, and backgrounds connected with Waiheke, tourism and wine (Baragwanath and Lewis 2009). The results all confirmed the problem of getting tourists to and from the cellar door so as to enable them to buy wine and then move on, especially in the winter when waiting or walking is far less pleasant or at the height of summer when buses are crowded and the wait for ferries and taxis is long.

The researchers identified the issue as one of *proximate isolation*, because even though Waiheke was close to Auckland, they were "isolated," in a fashion, due to their island setting. Visitor numbers were highly seasonal, and Waiheke wineries and enterprises therefore experienced two-fold isolation: first in attracting visitors to the island during inclement weather and off-season, and then in getting those that made it to the winery cellar doors.

The challenge was to find ways of securing and managing a flow of tourists to wineries across the year, as well as facilitating their access to the wineries at both ends. This included transportation from the ferry and around the wineries, and making sure the cellar doors and restaurants were open to accommodate them.

THE SOLUTION: RELATIONSHIP NETWORKS AND COLLABORATIVE MARKETING

Eventually the problem of *proximate isolation* rose high on the agenda of government agencies, individual wine enterprises, and the Waiheke Winegrowers' Association (WWA). Public meetings with tourism and ferry operators, winegrowers, and local public authorities were conducted, and the researchers were involved in analyzing meeting agendas and records. Over the course of several years, the following solutions were implemented to help solve the problem.

Relationship Building & Planning – The Waiheke Winegrowers Association (WWA) began building relationships with tourism enterprises on Waiheke, governance agencies, wine businesses in Auckland, and with Auckland Tourism, then the publicly-funded tourism agency responsible for promoting Auckland as a destination. The research study also generated interest from the ferry operators (Fullers and Sealink), Tourism Auckland (the wider regional tourism authority), and Auckland City Council, which had responsibility for roading, public transport, environmental management, and regulating expansion in winery operations. Chaired by David Irvine a principal in Cable Bay, the group connected up a new network of interests and discussed many of the issues at stake in Waiheke's wine tourism development.

Signage & Wine Map Developed - The WWA made efforts to engage their winery members to gear up for increased wine tourism by providing enhanced signage and encouraging members to formalise and extend visiting hours. They also produced a wine map identifying the locations and opening hours of wineries, and developed advertisements.

Welcome Center Established - The privately owned Waiheke Wine Center in the heart of Oneroa, the small township close to the ferry, became advertised as the retail outlet for purchasing local wines, encouraging and enabling visitors to experience Waiheke wines without the need for transport. In this way, tourists had a place to visit when the disembarked from the ferry, and could also obtain information about winery tours, restaurants, and other activities.

New Branding & Website – The stakeholders also found a solution in promoting and refining a collective brand: *Waiheke, Island of Wine.* At the heart of this branding solution lay the development of a new, more functional, more up-market website, which supported stronger links to wineries as well as new links to tourism providers. The website provides soft, collective infrastructure that has encouraged wineries to improve the quality of their own websites and communication with customers and tourists, as well as tighten up their tourism offerings. The website has allowed for tourist operators to improve the quality of their products. This collective infrastructure has recast the problem of *proximate isolation* as an opportunity to build a more effective wine tour proposition. It has helped to consolidate with the ferry companies and to encourage new initiatives among private wine tour operators.

New Wine Tours & Partnerships with Tour Operators– One of the most effective solutions to the problem of proximate isolation was the launch of a new suite of wine tours, which visit most of the commercially-focused wineries on the island. These wineries have in turn committed to keeping open cellar doors across the week and across the year. Wine tourists now have access to continuous and reliable cellar door experiences via wine tours offered by multiple providers, or by taxi, bus, rental car, bicycle, or foot. The new tours have particular features but tend to have a standard form: pick-ups from the ferry, visits to three wineries and associated tastings, a drop-off for lunch at a restaurant chosen by the tourist, with some time then for sight-seeing and return transport to the ferry. There exist tours that offer wider options, and cater for greater levels of independence or more intensely guided experiences. The tours are run both by the main ferry company, Fullers, and by private operators on the island. The car ferry operator, SeaLink also has a relationship with a tour operator.

Waiheke Wine Tour Alternatives, Summer of 2014-15

Tour	Features
Waiheke Vineyard Hopper	hop-on, hop-off bus service 6 stops / 8 wineries
Wine on Waiheke	Daily from Auckland Deluxe tour Tastings at three wineries
Around Waiheke Tours	Group bookings Twice daily half-day tours Wine as part of wider Waiheke experience
Wine on Waiheke	Four wineries 4-hour food and wine tour of Waiheke Island with round-trip ferry service from Auckland
Waiheke Island Wine Tours	Daily: three wineries; transfers to and from ferry and lunch Qualmark accreditation; TripAdvisor Certificates of Excellence Caters for corporate groups, specialized tours Includes option of deluxe two-day tour
Enjoy Waiheke (Waiheke Wine Tour)	Three or four vineyards Lunch at any location
Ananda Tours	Multiple options – base tour is pick-up from ferry, three wineries, drop off at lunch and pickup for return ferry
See Waiheke (Wine Lovers tour)	Visit four vineyards Boutique vineyards and award-winning wines (for the more discerning 'wine buffs')
New Zealand Wine Promotions	Part of a family of tours; pick up and drop off in Auckland; uses vehicle ferry (1 hr sailing); full-day, 2 person minimum booking 4 vineyards; Irregular bookings; Qualmark certified
Waiheke Premium Wine Tour	3 vineyards caters for particular audiences (e.g. Ladies Wine Tour)

Corporate Meetings & Weddings - The new, enhanced wine tour infrastructure also provides opportunities to foster winter tourism via corporate events, weddings, or specialized wine tours. Many of these tours seek to make the absence of a car a virtue rather than a constraint, especially with a tightening of New Zealand's drink-driving laws in 2014. They do not, however, necessarily displace more independent wine tourism, for those with access to transport.

Tourist Enjoying Wine on Waiheke Island

Conflict & Teamwork – It should be noted that over the course of the implementation of the many new solutions that some conflict occurred, which is only natural when so many diverse parties come together. Meeting records illustrate that some of the Waiheke wineries with different business models, resources, and pressing problems, and at different stages of enterprise development had markedly different commitments to collaboration and were divided on Waiheke's wine proposition. Some favored a more 'mass tourism' approach, while others sought an 'elite niche marketing' approach. Thus while the records demonstrate a nascent collective approach to issues, including access to cellar doors, the quality of experience and seasonality, this was far from uniformly accepted.

In the end however, there were many positive aspects to working together as a large team. As one Waiheke tour operator reported:

"Everyone has been represented singly up to this point and now they're working together to overcome the perception that it's too hard to get to Waiheke. The idea is that people can contact any of the 'team'

RESULTS AND WHY WAIHEKE ISLAND IS A BEST PRACTICE

The results of all of these efforts have had several positive effects on Waiheke Island wine tourism, which support this as a best practice. The number of tourists visiting Waiheke annually is difficult to determine because commercial sensitivities of the ferry operators means they choose not to publish this data. However a recent statement of evidence to the Environment Court suggests that the number of tourists has increased significantly since 2009. Winery owners and other tourism enterprises have noted these increases, which also reflect a similar increase in opportunity for further revenue generation. As one winery owner reported:

"The growth in numbers from organized wine tours and the increased professionalism of tour operators has allowed us to hire a staff member and organize our cellar activities around more consistent demand."

Furthermore, Waiheke Island has also become known as a leader both domestically and internationally in developing wine tourism (Baragwanath and Lewis 2014). Waiheke is increasingly promoted widely as the preeminent Auckland tourism destination. The increased professionalism and prominence of the tourism proposition has seen it integrated more fully into Auckland's tourism promotion, especially the increasing opportunities provided by cruise ships and increasing numbers of visitors generally and on cruise ships. Each of these collective responses has contributed to the increase of visitors to Waiheke, and the wineries are the beneficiaries. A recent tourist post on TripAdvisor supports this premise:

*"**6 Stars!!!** My favorite part of the trip to Australia and New Zealand. Such a beautiful place, just unbelievable. Six of us landed on island and rented a van for 89 NZ, which took us anywhere we wanted. Initially we're thinking about renting a bike there, but island was much bigger than we thought. Would highly recommend taste a wine there, even though I'm not a big fan of a wine, it feels*

totally different there. Atmosphere, sea, green all around, totally
delicious wine adds up. (Handaashka, Ulaanbaatar, Mongolia,
May 2015) "

While the relationships between the growth of wine tour packages, the new website, and enhanced relationships between tourism enterprises and Waiheke's wineries are difficult to establish, it is clear that they are connected and mutually reinforcing. Indeed, they each represent a related shift towards a greater level of professionalism in wine tourism. They also represent success through collaboration, albeit a collaboration centered as much on recognizing new opportunities from shared interest as enforced by collective action or codes of conduct.

The Island of Wine imagery, collectively developed by wineries, has been picked up by tourism marketing, leading to a collective improvement in the consistency of visitor traffic to Waiheke and the quality of experience. The wine tours clearly provide improved consistency of visits to wine enterprises with cellar doors and various hospitality enterprises, and benefit from increasing tourism to the island. This provides a solution in terms of a regularity of visits.

While the wintertime issues remain, the enterprises have sought to diversify their offering through corporate events, weddings, conferences, and corporate retreats. This diversification and associated collective marketing of the island is now increasingly led by tourism actors, including a new ferry operator that started operations in the summer of 2014-15, increasing access to the island.

FUTURE ISSUES

Despite the success of the relationship networks established and collaborative marketing efforts, Waiheke Island still has some issues to confront. One of these is how the wine tours have changed the experience of the tourist with the winery. The relationship is now with the tour operator and guide rather than the winemaker. This puts the quality proposition in the hands of tourism providers. Indeed, the tour becomes the product, which may mean that having pre-paid wine tourists no longer engage in the same way with the winery or feel a moral responsibility to buy wine or to connect with a winery in such a way as to promote it by word of mouth or buy its wine once home. As one of the small owner-operators observed, while structuring the commitment of time to the cellar door, wine tours still require a commitment of time and an interruption to vineyard and winemaking operations.

In addition, the arrival of a tour bus can also detract from the authenticity of the vineyard and cellar door experience sought by independent travellers or those seeking to enjoy lunch at a winery. Often the appeal of Waiheke's wineries is the environmental aesthetic and the social status of the experience. Tours can be invasive, especially on arrival, and disrupt these aesthetics, as well as distracting winery staff. They can crowd even the largest of Waiheke's cellar doors. These disruptions can be managed by timing tours appropriately, taking on a staff member to manage them, creating different spaces for dealing with them and so on; but this requires added management and often labor and other costs. Mudbrick, for example, has a cellar door and terrace café separated from its upmarket restaurant, and has different spaces for accommodating wine tours. However, it is a sizable hospitality enterprise, and even it can be overwhelmed by group tours that occupy the cellar staff and shop.

With access to the wineries provided by tours, the wineries face the on-going challenge of monitoring and sustaining quality experiences for both group and independent tourists, maintaining relations and negotiating terms with the tour companies, managing tour groups at the winery, and selling wine to tour groups so as to add value through the cellar door rather than just act as a tourism destination. There is a series of balances to be struck on an on-going basis. At the same time wineries must attend to seasonality, particularly to trying to stretch the on-season at each end. Wine tours have helped to address this problem, but there is likely to be a role for wineries in innovating new attractions, especially those that do not rely on sunshine. This will of course again raise questions about core business and business identity, especially for those high-end wineries and wine entrepreneurs for whom the vineyard lifestyle is important.

DISCUSSION QUESTIONS

1. How does seasonality complicate problems of proximity and access in building regional wine tourism?
2. What are the distinctive features of Waiheke as a wine tourism destination?

3. What types of agency have interests in wine tourism, and is it possible to develop responses to problems that align those interests?
4. In what ways can individual entrepreneurialism be harnessed to the work of industry bodies in promoting regional wine tourism?
5. How does the case illustrate the different ways in which academic research can contribute to generating solutions to applied problems in wine tourism?

Chapter Ten

Sonoma Sunshine: Learning to Collaborate for World Class Wine Tourism

Liz Thach & Michelle Mozell
Wine Business Institute, Sonoma State University, California

The vineyards gleamed golden in the afternoon sun of a late November day in Sonoma County, California. Vast and undulating, the tapestry of yellow and orange leaves of the grape vines seemed to flow across the horizon, climbing hills and wrapping around tall stands of dark green fir trees. Inside his office, Tim Zahner typed rapidly on his computer keyboard, oblivious to the natural beauty just a few miles from his Santa Rosa office. He had several critical emails he had to send, as well as a presentation he needed to complete on next year's tourism strategy.

As CMO of the Sonoma County Tourism Bureau for the past eight years, Tim's workload seemed to increase more each season, but now all of his hard work and the work of his partner organizations and staff appeared to be paying off. He paused in his typing to glance over at the plaque on the wall that read "Traveler's Choice Sonoma County #1 Wine Destination in the USA from TripAdvisor (2012)." A slow smile played across his face as he remembered how excited his employees were when they heard about the award.

Then he frowned quickly as he recalled the challenges of the early years. Those were tough times, when there were no collaborative partnerships between the associations promoting Sonoma County wine and tourism. It got so bad that the hotels were complaining about lack of customers and reviews on the TripAdvisor were generally more negative than good. The individual wine appellations (AVAs) were competing against one another, and there was no organized method to collect funding, let alone create a unified marketing campaign and track results. Even worse, Napa

Valley next door was being praised for their tourism efforts, while Sonoma County seemed to be languishing.

Tim leaned back in his chair and rolled his shoulders a couple of times to relax the muscles. Then he took several deep slow breaths and smiled again. That was all in the past, he told himself. Due to the visionary leadership of the Sonoma County Board of Supervisors and the formation of the "Trio," Sonoma County had become a top global wine tourism destination. In 2012 more than 7 million tourists visited to spend $1.55 billion (SCT, 2014), surpassing neighboring Napa Valley's tourism revenues of $1.4 billion in the same year. Now the new challenges were how to maintain the success and keep the image fresh so that tourists continued to pour into the region.

OVERVIEW OF THE AMERICAN WINE INDUSTRY

Historical sources state that the first US wine was made by French Huguenot immigrants in Florida in 1565 using the native American grape, Muscadine (Bates et al, 1989). Attempts to grow *vitus vinifera* grapes along the East Coast of America were generally unsuccessful due to the inability of European rootstock to withstand attacks by phylloxera and other pests. It wasn't until 1629, in the dry climate of New Mexico with its sandy soils, that *vitus vinifera* "Mission" grapes were planted by Spanish explorers and successfully crafted into wine (NMWGA, 2014). By 1769 when the Spanish started the first mission in San Diego, California, they had already planted many thriving vineyards in Old Mexico. The sunny dry climate of California was ideal for grapevines and soon vineyards were planted throughout the state, arriving in Sonoma County in the 1820's and Napa Valley in the 1830's (SCV, 2014, NVV, 2014).

# of Wineries in USA	8287
# of Wineries in Sonoma County	450

Today, wine is produced in all fifty US states, with the total number of wineries at 8,287 (Wines & Vines, 2014). Of these, 6,565 are produced by bonded wineries with a physical location, whereas the other 1,197 are created by virtual wineries. California currently accounts for a 47% of all US wineries with 3674 wineries in production. The next five states with the largest number of wineries are: Washington with 689, Oregon with 566,

New York with 320, Virginia with 223, and Texas with 208.

The US organizes their vineyard appellation system under the term "AVA," which stands for American Viticulture Area, and must be approved by the federal government. AVAs do not guarantee quality of wine but instead communicate an authentic and distinctive winegrowing region defined by specific geographical features and climate. Currently, there are over 220 AVAs in the US (TTB, 2014)

Map of Major US Wine Regions

The US is the fourth largest wine producer in the world behind Italy, France, and Spain, and has produced 23.6 million hectoliters of wine in 2013 (OIV, 2015). Of this, nearly 90% came from California (Wine Institute, 2014). Chardonnay grapes led the way in the 2013 California harvest at 16.1% percent, followed by Cabernet Sauvignon at 11.2%, and Zinfandel at 10% (USDA, 2014). Though the majority of US wine is sold and consumed in the country, in 2013 California exported 15% of its production for a total of $1.55 billion (Wine Institute, 2014).

Wine consumption has climbed steadily for the past 20 years in the US, reaching 2.82 gallons per capita, or 10.67 liters, in 2013 (Wine Institute, 2014). This increase has allowed the US to become the largest wine consuming market in the world at 329 million cases sold in 2013 for $36.3

billion (Impact Databank, 2014). Imported wine made up 34% of these sales. More than 80% of the wine sold in the US is less than $10 per bottle, with the average price around $9 per bottle (Brager, 2014).

The growth in US wine sales can be attributed to many factors, including an increase in the quality of American wines, increasing wine tourism, and the elimination of trade barriers within the global marketplace. In addition, the fact that wine has been featured more frequently in television shows and online, and that the younger Millennial generation has been adopting wine in large numbers, have helped wine to become more a part of American culture (Wine Market Council, 2013). Finally increasing retail outlets selling wine, as well as new online channels such as Amazon.com, have helped to bolster US wine sales.

Tourism and Wine in the USA

The US is the second most visited country in the world after France, with 74.8 million tourists arriving in 2014 (UNWTO, 2015). The category of "travel and tourism" delivered $2.1 trillion to the US economy in 2013, contributing to nearly 2.7 percent of the nation's gross domestic product, and creating 1 out of 9 non-farm jobs in the US (US Travel Association, 2014).

In California, tourism is an economic driver, as well, totaling $109.6 billion in 2013, with nearly 60% generated by international visitors and residents of other states (Discover California, 2014). Of this, wine tourism in California contributed $2.1 billion with 20.7 million tourists visiting California wine regions in 2013.

OVERVIEW OF SONOMA COUNTY WINE

Located in northern California about 40 miles north of San Francisco, Sonoma County is bordered by the Pacific Ocean on the west and Napa Valley on the east. The main city is Santa Rosa, and there are approximately half a million people living in the county (County of Sonoma, 2014).

In 1857, Agoston Haraszthy, a Hungarian native, established Buena Vista Winery outside the town of Sonoma. Today it is known as the oldest premium winery in California (McGinty, 1998). Following the development of Buena Vista, up until the time of Prohibition, many wineries that still exist today were established in Sonoma County, such as Foppiano, Korbel, Simi, Gundlach, Bundschu, and Sebastiani (SCW, 2014).

By 1920 there were 256 wineries in Sonoma County, with more than 22,000 acres (8900 hectares) in production (SCV, 2013). However, with the advent of Prohibition, when production of alcohol was made illegal in the US, many of Sonoma's wineries were forced to close and vineyards were turned over to other crops. Upon the repeal of Prohibition in 1933, fewer than 50 wineries in the county survived. Even as late as the 1960s, only 12,000 acres (4800 hectares) were planted to winegrapes (SCW, 2015). However, as American wine consumption grew, more and more vineyards were planted. As winemakers came to realize the potential to craft extraordinary, world-class wines from Sonoma-grown grapes, the number of wineries began to multiply as well.

Map of Sonoma County AVA's

Today, there are more than 450 wineries in Sonoma County and 64,000 acres (25,900 hectares) of vineyards (SCV, 2015; SCW, 2015). Many small multi-generational growers and winemakers operate in the region, with 40% of the vineyards comprised of less than 20 acres (8 hectares) and 80% less than 100 acres (40 hectares). Winegrape production remains, to a great degree, a family farming activity, with a strong commitment to sustainability (SCW, 2015).

Sonoma County is divided into 17 American Viticulture Areas (AVAs) that reflect the wide variety of climate and soil conditions in the region. The region produces 22 major grape varieties, with the three largest being chardonnay in first place, followed by pinot noir and cabernet sauvignon. In 2013, total tonnage of winegrapes harvested was 270,000 tons, valued at $605,068,400 (County of Sonoma, 2014). This averages $2249 per ton, and is the second highest value winegrape region in California behind Napa Valley at $3691 (Adams, 2014).

Wine Tourism in Sonoma County

In terms of tourism, the most recent survey (SCT, 2013), which is conducted every few years, showed that more than 7.5 million tourists visited Sonoma County in 2012 and spent $1.55 billion. Roughly 90% of the visitors were domestic with international tourists coming from Canada, Western Europe, Mexico, and Asia. On average, they spent $389 per day, with less than half on lodging. Tourism revenues provided total taxes to the government of $97 million, and created 19,000 jobs in Sonoma County. Interestingly, this is higher than Napa Valley tourism statistics in the same year, which show that 2.94 million visitors arrived, generating $1.4 billion in direct county spending (LNV, 2014).

THE PROBLEM: LACK OF COLLABORATION BETWEEN TOURISM STAKEHOLDERS

In the early 2000's Sonoma County was beset with a number of issues that made attracting tourists rather challenging. A primary one was the lack of a central tourism organization, which caused towns, visitor centers, and individual AVAs to focus only on their own areas. This resulted in no unified marketing message and little cooperation.

Another issue had to do with the old method of funding tourism in Sonoma County. Originally tourists had to pay a lodging tax, called the Transient Occupancy Tax (TOT), if they stayed at a hotel. The tax revenues were administered by the Sonoma County government, which dispersed funds to the individual visitor centers for marketing and administrative costs. Unfortunately only about 20% of the money was used for marketing, and this was applied towards the individual visitor center's town rather than the whole county.

Other problems were reports of poor or inconsistent customer service at local wineries, restaurants and hotels. Also, the fact that the very famous Napa Valley was a next door neighbor, and attracting more tourists than Sonoma County, made the situation more disheartening.

Sonoma County Vineyard with Mustard Flowers

A final issue had to do with lack of brand recognition for the name "Sonoma." Whereas Napa Valley had a regulation requiring that all of their wineries list Napa Valley on the wine label, Sonoma County did not. This resulted in many wineries in Sonoma listing the individual AVA on the label, such as Russian River or Alexander Valley, but not including the term "Sonoma." Therefore, when wine consumer surveys were conducted, Sonoma always scored lower than Napa in regional brand recognition.

THE SOLUTION: FORMATION OF SONOMA COUNTY TOURISM AND THE "TRIO"

The situation became so bad that a group of hotel owners, who were concerned over the lack of tourists, decided to work together to find a way to fund marketing for the region. The group approached Ben Stone, who was the Executive Director of the Sonoma County Economic Development Board (EDB). The EDB was responsible for assisting in the promotion of businesses within the county, and accountable for administering the funding for tourism as well as other programs. Ben began to investigate the situation and discovered that the visitor centers were also having issues.

So with the backing of the hoteliers, Ben began recruiting a group of leaders in the public sector and tourism industry to investigate ways of creating a funding structure to support tourism for all of Sonoma County. Through these discussions, the concept of a business improvement area (BIA) emerged. Finally in 2005, the Sonoma County Board of Supervisors established Sonoma County Tourism (SCT), which became the official destination marketing organization for all of Sonoma County.

Pacific Coastline of Sonoma County

Organizational Structure & Funding of SCT

Sonoma County Tourism (SCT) was established as a private, non-profit organization, with a mission to market and sell Sonoma County as a desirable destination to visitors who are traveling for leisure or business. It is staffed by twenty employees, and includes a CEO who oversees the three departments of marketing, sales, and finance.

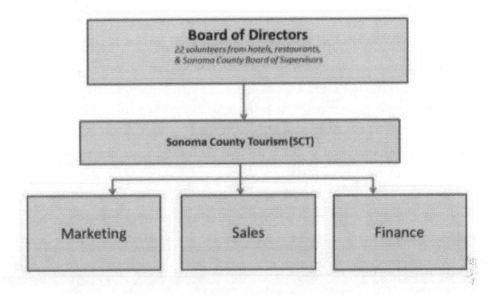

Organizational Structure of Sonoma County Tourism

SCT reports to a volunteer board of 22 directors. Thirteen members of the board are elected annually and the Sonoma County Board of Supervisors selects the other nine members. These 22 volunteers have full-time jobs as operators of hotels, restaurants, bed & breakfasts, and other tourist destinations within the County. Together they provide direction and oversight.

Funding for SCT comes from two sources. The first is a 2% assessment tourists must pay for lodging in Sonoma County and makes up about 66% of their budget. This money is considered as part of the Business Improvement Area (BIA) concept. The other 34% of the budget comes from the original TOT tax, which is also shared with individual towns, visitor centers, or for other purposes. The total annual budget for SCT ranges from $5 million to $6 million dollars.

Customer Segments of SCT

SCT focuses on three segments of tourists: 1) individual leisure travelers, 2) meetings and groups, and 3) tour operators. Its most important mandate is filling the county's hotel beds. The SCT marketing department therefore focuses on the specific needs of each of the three segments to encourage them to stay several nights or more in Sonoma County. They have actually created fictional characters so staff can better understand the needs and motivations of the different tourist segments.

For example, individual leisure travels come to Sonoma County for different reasons. Many come to taste wine, but others come to hike the trails along the ocean and through the redwoods; some come for food; and others come to get married in wine country. Therefore, Tim and his staff have developed some of the following characters: *Michael and Mary, the Boomer Man and Woman; Sienna the Millennial, Kim, the Foodie; Dan, the Gay Traveler, and Amy, the Bride.*

For the "meetings and groups," segment they have identified *Nancy, the Meeting Planner*, and *Wendy, the Wedding Planner*. For "tour operators," segment they have identified *Clyde, the Tour Operator*, and *Sandy, the Travel Writer*. By helping staff understand the specific needs of these various customer groups, they can create more impactful marketing programs, and encourage tourists to linger longer in the hotels, restaurants, and other sites of Sonoma County.

Operational Responsibilities of SCT

In order to promote Sonoma County, SCT employs a host of marketing tools:

1) **Website** - A sophisticated website in six languages where tourists can find everything they need in one place, including information on hotels, restaurants, wineries, tour companies, and even suggested tour routes throughout different parts of the county.

2) **Advertisements** - The placement of advertisements where SCT can maximize its exposure dollars: print, online, and broadcast. Examples include advertisements with *Sunset Magazine, WineSpectator*, and *Via*.

3) **Email Campaigns & Social Media** – placing online ads and interacting weekly or daily on eight social media channels: Facebook, Twitter, Instagram, Google Plus, Pinterest, YouTube, Flickr, and FourSquare.

4) **Press Releases** –about the organization and industry.

5) **Trade Shows** - where SCT "shares the booths and marketing efforts of trade partners and alliances."

6) **Partnering with National Tourism Groups** - such as VisitCalifornia (a nationally-focused organization), regional and municipal Chambers of Commerce, and the marketing groups of individual cities and towns.

7) **Partnering with International Tourism Groups** - Making multiple visits to overseas tourism-related offices to "shake hands," with groups such as Brand USA, which promotes US businesses overseas.

8) **Communicating with Tourism Partners** – Mailing monthly reports entitled *Tourism Update* and *Visitors Chronicle* to all partners, such as hotels, restaurants, wineries, etc. In addition, SCT holds an annual *Trends in Tourism* conference for partners, as well as "*Coffee Klatches*," which are informal meetings over coffee to brief partners on what SCT has accomplished to date. Also, SCT invites partners and local tourism businesses to host visiting journalists and trade, participate in SCT activities, and to add content to the SCT website by advertising events and updating listings.

9) **Partner with the "Trio"** (as described below) to support one another in promotional activities.

Formation of the Trio

Soon after the formation of SCT the concept of the 'Trio' arose. Since the wine industry is so important to Sonoma County tourism, it was critical for SCT to partner closely with the two local wine associations, namely *Sonoma County Vintners* (SCV) and *Sonoma County Winegrowers (SCW)*.

In the beginning, all three organizations were operating independently with separate funding, and in some cases were accidently duplicating efforts. When SCT was formed, the three organizations started working more closely together, and, when an opportunity arose for them to share office space and administrative staff, they agreed that it made sense from a cost-saving perspective. Therefore, all three organizations moved into the same building in Santa Rosa and began working more closely together. Consequently, they not only helped save money for the county, but also began to collaborate more effectively in communicating with tourists and partners, and streamlined their activities. They began to jointly plan activities and work together as a "trio team" to support Sonoma County.

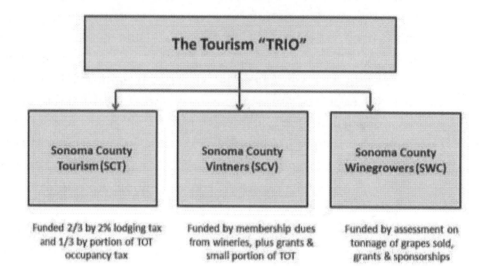

Diagram of Sonoma County Trio Collaboration

Sonoma County Vintners (SCV)

Sonoma County Vintners (SCV) is the oldest of the three agencies within the "Trio," founded in 1944. Their mission is to "establish Sonoma County as the leading winegrowing region recognized globally for superior wine quality, varietal diversity, unparalleled scenic beauty, and culinary excellence." (SCV, 2014, p. 2) Currently, they represent more than 250 wineries and affiliates of all sizes throughout the county, with a seven member team to implement the following:

1) **Website:** to communicate Sonoma County wineries(sonomacounty.com)
2) **Print Advertising:** in "food and wine" magazines.
3) **Online Promotion**: Google ad placement to drive traffic.
4) **Email communication, Press Releases, and Social Media**: Facebook, Twitter, and Youtube
5) **Trade Events:** to conduct tastings and presentations at high-level trade events both nationally and internationally, such as Sonoma in the City and the Hong Kong Expo.
6) **Special Local Events:** such as the 3 day Sonoma Wine Country Weekend which attracts more than 5,000 visitors each year; the

Sonoma Harvest Wine Auction that raised more than $4 million for charity in 2014; and the Sonoma Summit that brings in 35 top sommeliers and wine buyers.

7) **Partnership with Visa Signature Credit Card:** SCV works with Visa to provide winery discounts for visitors using a Visa Signature card, as well as setting up winemaker dinners around the country, where funding is shared with Visa.

8) **Partner with the "Trio":** to support one another in promotional activities. For example, SVC asks their member wineries to donate wine that can be poured at SCT and SCW events.

Sonoma County Vintners is a membership-funded organization from which the bulk of annual revenue is derived. Membership dues are based upon a winery's annual case production: a winery producing under 6,000 cases a year pays a flat fee of $1,000 per year while larger wineries (over 6,000 cases a year) pay $1,000 plus 10¢ per case over 6,000 cases per year (with a cap at $10,000). They also receive a small amount of the TOT funds from the County of Sonoma, just as Sonoma County Tourism does, and occasionally receive federal agricultural grant money.

Sonoma County Winegrowers (SCW)

The mission of Sonoma County Winegrowers (SCW) is to "increase the value of Sonoma County winegrapes and to nurture and protect this agricultural resource for future generations." (SCW, 2014). It was initially established by grape growers in Sonoma County in the 1980's as a member based trade association. Formerly called the Sonoma County Winegrowers Commission, it is a marketing and education organization that represents grape growers, and has a staff of eight people.

1. **Marketing/Communications** - promote the organization and growers through national advertising and public relations via their websites (sonomawinegrape.org and wearesonomacounty.com), social media, and events, such as Grape Camp for consumers, Sonoma Summit for Sommeliers and Wine Trade and Sonoma in the City for trade, media, and consumers.

2. **Grower Education Programs** – provide a series of educational workshops, seminars, trade shows, and events throughout the year for grape growers, including sessions on farming practices, pest management, vineyard management, employee development, business and marketing, and events such as a county-wide pruning championship. Some of these programs are in Spanish.

3. **Grower Resources** – provide information and reports on vineyard practices, updates on permitting, regulations, labor, grape prices, and other critical issues. SCW also has a Grape Marketplace to assist growers in finding buyers for their grapes.

4. **AVA Grants** – provide grants to support individual AVA's in marketing themselves and to support their participation in SCV and SCW hosted events.

5. **100% Sustainable Wine Region Goal** - provide support for growers to enable Sonoma County to achieve its commitment to become the nation's first 100% sustainable wine region by 2019. Part of this role includes increased education on sustainability best management practices, as well as assisting growers in completing the *California Sustainable Winegrowing* assessment and/or certification program.

6. **Partner with the "Trio"** to support one another in promotional activities - For example, SCW partners with SCT and SCV to host events that market Sonoma County as a destination by featuring Sonoma County wines, winemakers, and grape growers.

SCW is funded by an assessment on grape production of .5% of the per/ton price receipt. This was voted upon by the growers in 2006 when the commission was established, and provides the majority of the funding, which is around $1.2 million per year currently. The assessment is collected through the wineries in their contract with the grape growers. A secondary source of funding is from grants, such as the Specialty Crop Block Grant, which is a two-year grant of federally-supplied funds designated for each state to use toward the support of agriculture, including marketing. These grants are especially important during low production years. A final source of funding comes from sponsors. SCW has five levels of sponsorship, ranging from $350 to $10,000 per year.

Development of Sonoma County Logo and Conjunctive Labeling

One of the major outcomes of the collaboration of the Trio was to share the services of a public relations firm. Since they all worked together with the firm, they were able to save money and also agree on a common logo. Though it took many long months of discussion and multiple designs, the three organizations finally agreed on the "Sonoma County Brand Stamp," which is modeled on an old-fashioned brand like a postage stamp or a brand on a cow.

The brand is used on all three organization websites, as well as on all promotional material, brochures, wine maps, and advertisements. It creates a unifying message for consumers about Sonoma County.

After agreeing on a common brand, the Trio next tackled the issue of consumer confusion over lack of consistent branding of Sonoma County wines. Whereas Napa Valley Vintners and members had wisely agreed years ago to require that all wineries include Napa Valley on the label if the grapes were from Napa Valley and the wine was produced there, Sonoma County had not done this. Therefore, many Sonoma County wineries only included their individual AVA, such as Russian River, and did not put Sonoma County on the label.

Sonoma County Brand Logo

The first step the Trio took was to conduct a consumer survey that illustrated how confused consumers were concerning AVAs in Sonoma County. The results showed that many consumers recognized Russian River

Valley more often than Sonoma County, and some did not know that Russian River was actually an AVA within Sonoma County (Johnson & Bruwer, 2007).

This data was shared with all Sonoma County wineries, and after months of discussion, including working with the TTB (Tax and Trade Bureau – a federal organization that must approve all wine labels) and ABC (Alcohol Beverage Control) Board of California, a conjunctive labeling law was passed.

According to the SCV website, conjunctive labeling is defined as "labeling of a wine to show both region and sub-region (AVA) of origin. In our case, it refers to the inclusion of "Sonoma County" on the label of all Sonoma County wines along with any AVA designation." There are three objectives for doing this:

1) To build brand equity for Sonoma County wines and preserve and strengthen Sonoma County's position as a recognized world-class wine region.
2) To increase sales of wines produced from Sonoma County grapes.
3) To increase recognition for every AVA within Sonoma County, both well-known and less familiar, and ensure that consumers understand where they are.

The law went into effect on January 1, 2011, but wineries had three years to comply so as to make sure to include "Sonoma County" on the label if the wine was produced from a Sonoma County AVA. The wineries have creative flexibility on the font, size, and location of "Sonoma County" within their label design. Labels must be submitted to the TTB for approval, and the conjunctive labeling law is enforced by the ABC.

Establishment of the Certified Tourism Ambassador Program

The next step was for the trio to tackle the issue of inconsistent customer service at Sonoma County hotels, restaurants, and wineries. With the rise of social media in the last half of the 2000's, a worrying number of tourists began publishing negative reviews about service. Fortunately the Trio responded quickly and developed a solution in the establishment of the Certified Tourism Ambassador (CTA).

The CTA program is an international certification program designed to train front-line employees and volunteers in the hospitality industry (ctanetwork.com, 2014). The goal is to improve visitation by inspiring front-line hospitality employees and volunteers to work together to turn every visitor encounter into a positive experience (SCT, 2014). Each participant attends a four-hour tourism training session and then completes a certification test. The course work includes learning Sonoma County facts, consumer service tips, and how to be friendly and positive. Sessions are held at various locations throughout Sonoma County, including many wineries who donate space for the training. There is also a certification renewal option. Graduates of the program receive a gold pin to wear on their clothing.

To date, over 1000 Sonoma County tourism-related professionals have completed the certification. The results appear to be positive based on many affirmative reviews on social media sites, including the following example:

> *"I recently had the most relaxing getaway of my life. It was that good. A big part of it was the chance to spend time with friends on a girls-only getaway. Another part of it was our chosen activities. We spent the weekend wine tasting in Sonoma. We were there with the express purpose of relaxing.*
>
> *Part of our success in turning off the intensity and getting relaxed can be attributed to the people we encountered. They were relaxed. And they were service-oriented. They helped us ease into our four-day getaway and to kick back, knowing everything would be taken care of for us. I started calling it Sonoma-Style Service because it was apparent almost everywhere we went in Sonoma."* - (Excerpt from July 10, 2013 Sonoma-Style Service)

RESULTS AND BEST PRACTICE IMPLICATIONS

The results of the establishment of the Sonoma County Tourism Board and the collaborative efforts of the Trio are well documented in the number of increased visitors staying in hotels as well as the Tourist Occupancy Tax revenues. According to Tim Zahner, "occupancy levels in Sonoma County lodging increased from 62% in 2005 to 76% in 2014, and TOT revenues increased from $16 million in 2005 to $27 million in 2014. We are quite happy with these results."

This example of tourism in Sonoma County can be considered a global best practice not only because of these impressive increases in visitors, revenues, and wine-related activities, but because of the extraordinary collaboration between government and non-profit organizations that does not occur that frequently.

The implications are very positive for Sonoma County residents and tourists, but also reflect well on California wine tourism in general. Furthermore, Sonoma County and Napa Valley appear to be much more supportive of one another with both counties encouraging tourists to also visit the other, as well as the neighboring wine destinations of Mendocino, Lake, and Solano counties. Indeed, an employee of the SCV stated, "We are good for one another. We need and support each other (Thach, 2014, p. 15)."

The benefits of working together in a collaborative fashion, as the Trio does, is summed up well by Sara Cumming, Director of Communications for Sonoma County Vintners (SCV):

"We've come to realize that the 'Trio' and working together has been super powerful. As wine perception goes up, grape production goes up. When 'wine people' come here to appreciate our wines, they stay in our hotels, so 'tourism' cares. We are all in this together. A lot of regions don't work that way. Political boundaries, egos, whatever - they don't see the synergy of it, so it puts us at an advantage."

FUTURE ISSUES

Though Sonoma County has accomplished much in the past decade through collaboration and increased tourism, there are still some major challenges the Trio recognizes:

- ***Sustaining the Growth*** – is a common concern with all three organizations. According to Tim Zahner, "We want to grow in a way that is sustainable and does not take on new costs." At the same time, they are continually trying to improve as well as implement new tourism features and events. For example, SCV was able to create a new event for Robert Parker the past year where he tasted more than 650 Sonoma County wines. SCV would also like to grow the Sonoma County Wine Weekend to be as large as the Aspen Food and Wine Festival. New tourism features, such as Sonoma County Canopy Tour, a zip line through the redwoods, has been attracting more types of tourists, and they want to continue to expand and grow in this way.

- *Funding* – continues to be a challenge, especially since all three organizations rely on different funding mechanisms. The number of tourists who stay overnight in Sonoma County hotels determines funding for SCT, whereas membership provides the majority of the funding for SCV, and grape prices and tonnage each harvest drives the budget for SCW. Fluctuations in any of these funding mechanisms can cause problems for all three of the organizations.

- *Measurement* - the way success is measured and how it ties to funding can also be a challenge. According to Tim Zahner, "Our funding measure is based on where tourists sleep in terms of county lines. Is it Sonoma, Napa, Marin? In a way, this is a trap. Somewhere in the future we'll figure out how we manage as a complete region."

In conclusion, the collaborative efforts of the Sonoma County Trio can be seen as a best practice and example for other wine regions around the world. It also illustrates a team spirit, a focus on community, and a love for the land that is critical to successful wine tourism. Karissa Kruse, President of SCW, sums this up eloquently:

"My personal vision is that Sonoma County continues to be seen as a place to come and explore. That it is about the people here: our growers, our winemakers, our hotel managers and staff, who are all trying to create a great experience and who also live here. They are 'walking the talk.' This is their community too. They feel a special bond and an emotional connection to this place. I want the consumer to feel this as well."

DISCUSSION QUESTIONS

1. Sonoma County has done a good job marketing itself as a major wine tourist destination in the US, and to some extent internationally. What steps can it take to improve its image and recognition on a global basis?

2. Assume you were assigned to a taskforce to examine the current funding mechanisms for tourism in Sonoma County. What recommendations would you make?
3. Conduct a SWOT Analysis on the Sonoma County Trio. First identify a list of Strengths, Weaknesses, Opportunities, and Threats. Then develop an action plan on what they should do moving forward to sustain their success.

Individual Winery Best Practices

"Accept what life offers you and try to drink from every cup. All wines should be tasted; some should only be sipped, but with others, drink the whole bottle."

— Paulo Coelho, Brazilian author of <u>Brida</u>

Chapter Eleven

From Piping Water to Piping Wine: The Zuccardi Wine Dynasty of Mendoza, Argentina

Jimena Estrella Orrego & Alejandro Gennari
National University of Cuyo, Argentina

On a sunny day in March, Julia Zuccardi led a group of tourists from the winery tasting room to a large vineyard outside. The verdant green vines were trained high on the "parral" trellis system, which was brought to Argentina from Italy. Large clusters of purple grapes hung from the vines, and a small sign in front of the vineyard announced it was organic malbec.

"My grandfather planted this vineyard," said Julia, Director of Hospitality for Zuccardi Winery. "He was a visionary."

She then told the visitors the story of the Zuccardi wine dynasty and how when her grandfather, Alberto Zuccardi, came to the Maipú Valley located outside the town of Mendoza, it was only a dry arid zone. There was no agriculture, no irrigation, and absolutely no tourism.

Today, more than 50 years after Alberto arrived, the Zuccardi vineyards are vigorous and healthy, and their wines are sold around the world. They own more than 1000 hectares of land, produce around 20 million liters of wine per year (over 2 million cases), and operate two award-winning restaurants. They also employ over 400 people, and were the first in the area to open a hospitality center for wine tourists. Now they have over 40,000 visitors per year, and the winery is one of the few in the Mendoza region that is open seven days a week to welcome wine tourists.

Zuccardi Winery is considered to be a best practice in wine tourism in Argentina. This chapter explores the history and steps the Zuccardi Dynasty took to achieve such a successful operation, as well as the challenges they had to overcome, and future issues they may be facing.

OVERVIEW OF THE ARGENTINA WINE INDUSTRY

The origins of the Argentinean wine industry can be traced to the Spanish colonial period and the establishment of the Virreinato del Rio de la Plata. However, the emergence of the modern wine industry is explained by the Italian and Spanish immigration of the late 19th century and by the contribution of European specialists hired by the emergent Schools of Agriculture, such as the Faculty of Agrarian Science from the National University of Cuyo.

The planting of French varieties (cabernet sauvignon, merlot, tannat, and especially malbec), Italian varieties (barbera, nebbiolo, sangiovese, and bonarda) and Spanish ones (tempranillo, semillón, and pedro jiménez) took place simultaneously with the introduction of the railway in the provinces of Mendoza and San Juan, thus favoring immigration and the diffusion of new growing and winemaking techniques.

The irrigation system and water management organization also began at the time, giving a key contribution to the wine industry emergence. Large family wineries of Italian (Giol, Gargantini, Tittarelli, Cavagnaro, Filippini, Rutini, etc) and Spanish (Escorihuela, Arizu, Goyenechea, etc) origins were born, grew, and consolidated in this period (Mateu, 2008).

Today Argentina has over 223 thousands hectares of wine grapes, 1301 wineries, and ranks 5th in the world in wine production (INV, 2015). Its signature grapes are malbec for red wine and torrentes for white wine.

Argentina's wine production areas range from the northern province of Salta to the southern region of Patagonia. This strip is characterized by aridity and dryness and is irrigated by melted water from the Andes, creating oases of green. These oases can be classified into regions and sub regions, each of them with particular characteristics in terms of geomorphological conditions. Some stand out for their altitude, such as the Calchaquíes Valleys, in the North; others for the aridity of the land, such as the valleys in the provinces of Mendoza, San Juan, and La Rioja; and there are also low altitude oases in Patagonia, with intense ripening periods.

Three Major Winegrowing Regions of Argentina

1) **Cuyo Region:** The Cuyo Region, named after the native Huarpe word for "the land of deserts", is the central wine producing area in Argentina. It includes the provinces of Mendoza, San Juan,

and La Rioja. Topographically, it is composed by rugged mountainous relief. With the snow-covered Andes Mountains towering over the vineyards planted on sandy desert-like land. Vineyards ranged in altitude from 700 to 1700 meters above sea level.

Map of Argentina Wine Regions

Mendoza is the most important among the wine producing provinces in Argentina. It represents more than the 80% of all the wine production in the country, and is composed of five sub-regions: North, East, Center, South, and Uco Valley. The main grape varieties of the province are: malbec, merlot, cabernet sauvignon, torrontes, chardonnay, sauvignon blanc, and viognier.

San Juan is the second largest wine producing area in Argentina. Several valleys run through it, with Tulum, its most important area, sitting along the banks of the San Juan River. The main grape varieties of San Juan are syrah, malbec, cabernet sauvignon, bonarda, chardonnay, and torrontés.

In La Rioja the most important wine producing area is the Famatina Valley, located between the Velasco and Famatina Hills. Torrontés riojano is the typical variety of the region.

2) North Region: The Northern region includes the provinces of Salta, Tucuman, and Catamarca. It is known as the location of the highest altitude vineyards in the world (Decanter, 2014), ranging from 1,000 meters and 2,900 meters above sea level. The village of Cafayat, located in the Calchaqui Valley of the Salta region, is considered to be one of the most charming wine tourist villages in the country. Much of the best torrontes produced in Argentina comes from Salta, with Catamarca and Tucuman as emerging wine areas, where torrontes is also cultivated.

3) Patagonia Region: Patagonia is the southernmost region of Argentina where grapes are grown. It covers the provinces of Neuquén, Río Negro, and La Pampa. In the Neuquen province, the main wine valley is San Patricio del Chanar where sauvignon blanc, merlot, pinot noir, and malbec are planted. In Rio Negro, Alto Valle is the main area for wine production and it shares many characteristics with the Neuquen province. La Pampa is known for merlot, malbec, cabernet sauvignon, and chardonnay. During the past years, wine production has also extended to non-traditional wine areas such as Buenos Aires, Cordoba, Entre Rios, and Jujuy.

Wine Tourism in Argentina

According to the World Tourism Organization (UNWTO, 2015), Argentina was ranked first in South America for the highest number of tourists at 5.9 million in 2014. This is most likely because Argentina has been actively promoting tourism since the 2001 currency devaluation. Before this economic crisis, the country was quite expensive for tourists and almost no inflow was registered. When the currency devaluation occurred Argentina became an attractive destination for tourists around the world.

Wine tourism is also actively promoted via the organization Bodegas de Argentina, which took the necessary steps to position Argentina in the one touristic network. This paid off when Mendoza became one the Great Wine Capitals in 2005. More recently, *Wine Enthusiast Magazine* (2014) ranked Mendoza in the top ten wine destinations to visit. The region is famous for the annual Vendemia Wine Festival that lasts 10 days and is one of the largest in the world.

# of Wineries in Argentina	1301
# of Wineries in Mendoza	928

In 2010, the National University of Cuyo and the Fondo Vitivinicola organization conducted a study and determined the economic impact of wine tourism was six thousand million Argentinean pesos for Argentina. Currently estimates indicate that more than 1.5 million tourists visit Mendoza wineries each year (Thach, 2014).

ABOUT THE MENDOZA REGION AND ZUCCARDI WINERY

The Mendoza wine region is the largest in Argentina in terms of production. It is located along the foothills of the Andes with a continental climate that is hot in the summer and cool in the winter. Rainfall is low, so traditionally, irrigation channels funnelling water from the snow-clad mountains fed the vineyards, but more modern vineyards have adopted drip irrigation. The warm climate is ideal for producing big velvety red wines, and the region has become world-famous for Malbec wine. Today Mendoza has approximately 928 wineries (INV, 2015).

Mendoza is divided into different sub regions, with the areas of Lujan and the Uco Valley, south of the city of Mendoza, the most well-known to tourists. However, when Alberto Zuccardi arrived in Mendoza in 1963, he chose to settle in the unknown and undeveloped region of east Maipú Valley, located about 38 kilometers, or a 40 minute drive from downtown Mendoza.

Alberto had been raised by his Italian parents who had emigrated from Avellino, Italy, and settled in the Tucuman region of Argentina, 12 hours to the north of Mendoza. After studying to become an engineer Alberto became fascinated with designing irrigation systems using special cement he had developed. In an effort to sell his invention to farmers, he decided to set up a demonstration system in the very dry and arid region of east Maipú Valley. Therefore, he planted a vineyard as part of his demonstration, and began piping the cement needed to build the irrigation system.

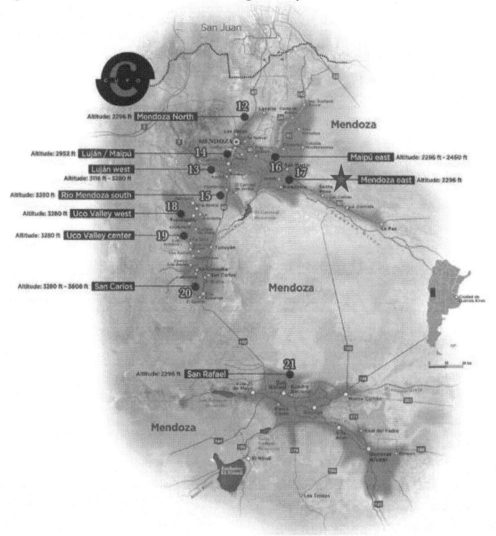

Map of Mendoza Wine Region

Though his original intention was not to start a winery, Albert soon developed a passion for grapes and wine. With his irrigation business a big success, he and his wife Emma began planting more vineyards and in 1968 started to construct a winery. In the beginning, they focused on producing table wine, which was only sold in Argentina. However, as their son, José Alberto, grew older and assumed a leadership position in the company, he made a strategic decision to shift towards producing high quality wines.

Today, José Alberto's three children, Sebastian, Julia, and Miguel, are involved in running the winery. With a focus on the four goals of high quality wine, innovation, environmental harmony, and giving back to the community, the three generations of Zuccardi's have managed to propel their winery to one of the most well known in Argentina. They have won numerous awards not only for their wines, but also for their business practices.

Sebastian, Miguel, Jose Alberto, and Alberto Zuccardi (left to right)

Familia Zuccardi is now one of the largest family-owned wineries in Argentina and maintains its reputation for quality in all markets by making and selling wines made solely from the family's own grapes. The company is run as a family, by a family, and is committed to doing things manually and carefully, keeping the use of machinery in the vineyards to a minimum.

The Familia Zuccardi team consists of 440 permanent staff with additional people during harvest. In many cases, whole families are employed – many of which are the descendants of those who worked in the vineyards when they were first planted. For the Zuccardi Family, "good people in healthy vineyards create good wines," and therefore, they confer as much responsibility as possible on the families and teams who work the vines.

They farm more than 800 hectares of vineyards, but not only in Maipu. They have purchased vineyards in the Uco Valley in the districts of Vista Flores, Altamira, and La Consulta, as well as in the Santa Rosa area. Almost 35% of these vineyards are certified organic, and in the rest of them sustainable production systems are used.

Altogether, they produce 30 different wine grape varieties including the white grapes of chardonnay, torrontés, viognier, chenin blanc, sauvignon blanc, pinot grigio, and pinot bianco. Red grapes include cabernet sauvignon, malbec, tempranillo, bonarda, merlot, sangiovese, syrah, pinot noir, caladoc, tannat, ancellota, barbera, marselan, marsanne, allianico, grenache, gamay, mourvèdre, and bourboulenc.

But achieving all of this was not very easy in the beginning. The Zuccardi Family faced many challenges, but the largest of these was attracting tourists to their unknown region of Mendoza.

Julia Zuccardi in Vineyard with Parral Trellis System

THE PROBLEM – NON-DEVELOPED TOURIST AREA

By the late 1990's the Zuccardi wine brand was selling well in Argentina, especially the very popular Santa Julia line, created for Alberto's grand-daughter, Julia. In 1999, the Zuccardi Q brand was introduced. This was their highest quality wine, with the Q standing for "Quality." The wine was well received, especially the Tempranillo, which was the first ultra-premium wine produced in Argentina from this varietal.

Despite the success in wine quality and branding, the Zuccardi's were concerned because there were very few visitors who took the time to make the 40-minute drive from Mendoza to their winery. While Mendoza was registering an increasing flow of national and foreign tourists, most of these tourists visited the Lujan and Uco Valley regions, ignoring the eastern part of the province where Zuccardi winery was located.

Even though Zuccardi offered wine-tastings and tours of the cellar as other operations did, they received very little attention. Therefore with their strong value of innovation, they decided to create a wine tourism experience that was completely different from what other wineries were offering. With Julia at the helm as hospitality director, they developed the following mission for their customer: "To endeavor to make your visit more than a tour through the tanks and barrel rooms; we want to make it an unforgettable experience." (Zuccardi.com,).

THE SOLUTION – CREATION OF A COMPREHENSIVE WINE TOURISM COMPLEX

Convinced that tourism was about creating moments and experiences, the Zuccardi family worked hard to develop an offering that was unique and memorable. In order to do this, they first conducted some research, and then over the course of several years introduced the following concepts to the winery in order to create a comprehensive wine tourism complex, which they call "Casa del Visitante."

Research on Wine Tourism Gaps

To begin they researched what competitor wineries were offering in the Mendoza area as well as wine tourism experiences in other parts of the world, especially California's Napa Valley. Through this process, they discovered two findings:

1) Most other wineries in Mendoza required advance reservations to visit and were not open on Sundays.
2) Though other local wineries offered tours, tastings, and restaurant facilities, they did not offer a portfolio of wine experience similar to those found in other parts of the world. For example, other wine regions offered multiple activities, such as cooking classes, participation in pruning and harvesting, hiking tours, balloon rides, vehicle rides through the vineyards, and many other unique wine tourism experiences.

Expansion of the Visitor Center - Open Every Day

The first change the Zuccardi's implemented was to adjust the opening hours of their winery to seven days a week. Visitors could drop by anytime between 9 am and 5:30 pm (10 am to 4 pm Sundays) for a tour and tasting with no advance reservations. This policy alone set them apart from other wineries in the Mendoza region.

Signage with Visiting Hours at Zuccardi Winery

The expansion of the visitor center included opening a wine shop, featuring not only Zuccardi wines, but other related products. These included oil olive and cans of olives produced on the estate, wine-based creams and soaps, cookbooks on Argentinean food recipes, wine maps, and t-shirts. In addition, a private tasting room was added to the second floor of the visitor center. It includes pieces of art specially chosen by Alberto, and a relaxed setting where tourists can enjoy intimate tastings and private lessons on wine.

Addition of Art Gallery and Olive Oil Center

Another tourist attraction was the developed of an art gallery, featuring both local and foreign artists. Each month a new exhibit is mounted, and special wine tastings are organized to complement the art. In this way, visitors can taste Zuccardi wines while enjoying paintings and sculptures.

Because the Zuccardi estate includes 400 hectares of olive trees planted by Miguel Zuccardi in 2004, they decided to make this part of the tourist initiative. Now, with a specialized tourist guide, visitors can harvest the olive, and participate in the complete production process of oil extraction, bottling, and olive oil tasting. Then they get to take home a bottle of their own olive oil. With distinguished flavors and aromas, the family produces over 300,000 liters of olive oil, including Frantoio, Changlot, and Arauco.

Launch of Two Restaurant Venues

In 2005, the Zuccardi's opened the first of two restaurants. Also named "Casa del Visitante," this restaurant is situated in the middle of vineyards and olive trees gardens, and focuses on offering traditional Argentine meats and BBQ, called "asados." There are two menu options at the restaurant: a traditional menu features key Argentine fare such as empanadas, a barbecue selection, salads, and grilled vegetables; and the tasting menu, which is a gourmet option presenting regional and seasonal products. Every course features a different Zuccardi wine to match the cuisine.

In 2013, the second restaurant "Pan & Oliva" was launched. The concept was to create a more relaxed family oriented venue with the menu based on different tapas, salads, and pastas. The majority of the fruit and vegetables are produced in their organic garden. The décor is similar to an old style Italian restaurant with red and white checked tablecloths and both indoor and outdoor seating.

Interior of Casa del Visitante Restaurant

Multiple Wine Tourism Experiences

Over the years in her role as hospitality director, Julia Zuccardi has expanded the number of tourism experiences. With a staff of 30 employees, she has added the following fun and unique activities for visitors:

- Biking and wine tasting
- Four track and wine tasting: Fourtrack ride from the gardens of Casa del Visitante to the vineyards of Finca Beltrán accompanied by a guide and tasting different wines in each of the three stops.
- Hot air balloon ride and wine tasting
- Horseback riding and wine tasting
- Cooking classes for adults
- Cooking classes for children
- Bartender courses for the many unique wine cocktails they have developed
- Picnic in the gardens
- Afternoon tea
- Harvesting grapes
- Pruning grapes
- Making olive oil
- Wine tasting classes
- Olive oil tasting classes

For the more active tourist, harvesting and pruning with winery guidance is offered. Special activities are also designed for Father's and Mother's Day.

Partnership with Tour Operators and Billboards

Due to forty minute drive from center of Mendoza to the winery, the Zuccardi's established partnership with travel agencies and tour operators. These include negotiating special rates, and providing them with Zuccardi Winery marketing material. On a regular base, the winery hospitality team visits local hotels to update them on their products, and invites hotel concierges to the winery for a special tasting and tour. For independent tourists who chose to drive to the winery, they have created a friendly website with directions, and have installed many billboards along the roads to help entice and direct them to the winery.

Strong Emphasis on Environment and Water Conservation

The Zuccardi family prides itself on environmental leadership and water conservation. The implementation of organic farming methods in some of their vineyards as well as the garden is a testament to this.

Flood Irrigation System in Vineyard with Parral Trellis

In terms of water conservation, José Alberto still uses the historical flood system because he believes it allows the parrals (high trellis system) to develop their own micro-climate in a more efficient way than could be achieved with drip irrigation. He also believes that the flood system allows the nutrients in the soil to be spread more evenly across his vineyards -- and further, that flood irrigation could well be the reason phylloxera has never struck Argentina's vineyards.

With the flood system, shallow walls of earth are constructed every fourth row, at 12- or 15- meter intervals. Irrigation is used only when necessary to control evaporation levels and to reduce potentially harmful stress on the vines.

Putting his keen interest in science and his innovative spirit to work, José currently is experimenting with three distinct irrigation methods in his Maipú vineyards. One operates at 30.4 centimeters (12 inches) above the ground, while the other is attached to the parral's overhead wiring, producing a rain-like effect between the vines. The third trial area uses a new micro-spray system.

Innovative Practices

Innovation is a key value for Familia Zuccardi. By focusing on research and development, the family is working hard to further understand their vineyards and increase winemaking quality even more. For example, they experiment with new grape varieties, such as verdelho, caladoc, and ancelotta that have not previously been planted in the region. They have created a small, experimental winery with twenty stainless steel tanks to test different vinification methods. Most of these special small-batch wines are sold in the winery shop, allowing immediate feedback from national and foreign consumers. Successful experiments have defined the first viognier from Argentina, new line of wines for the Canadian market called Fuzion, and the creation of a fortified malbec called Malamado. The term "*Malamado*" has a double meaning of "mablec made like Port" and "bad love".

Co-fermentation is another interesting project of the R&D team. By co-fermenting red and white grapes they found the resulting wine gained in length and complexity, especially for high tannin varieties. Today one of their best co-fermentations is with malbec, cabernet sauvignon, cabernet franc, and a touch of torrontes.

New experimentation is taking place in concrete tanks with a conical egg-shaped head. The objective is to verify if this "more natural" system offers special characteristics to the wines, such as complexity and roundness. Naturally, the concrete they are using for the tanks is from their own concrete business. They have built a new high-end winery in the Uco Valley that is fully equipped with these concrete tanks.

Zuccardi has also innovated strongly in the sparkling wine world. In 2004, winemaker, Sebastian Zuccardi, created Alma 4, a high-end sparkling wine. Soon after he developed a red sparkling wine made from the bonarda grape. Now, they have a wide portfolio of non-traditional sparkling wines.

Another aspect of the innovative nature of Zuccardi is their leadership in the areas of organic and sustainable farming. After many years of working on the protection of air, soil, flora, and fauna, they are now one of the largest organic producers in Argentina. While the entire estate follows sustainable agricultural practices, half of the vineyards are certified organic. The organic varieties being produced are: malbec, sangiovese, chardonnay, tempranillo, cabernet sauvignon, torrontes, and bonarda. Their wine brands, Organiz Fuzion and Santa Julia Organic, are some of the top sellers in North America.

Marketing Communication and Collaboration

Zuccardi employs a sophisticated marketing strategy to entice tourists to the winery. This includes a focus on collaboration with tourist agencies and hotels, and is implemented via traditional and digital promotions. For example, they place multiple print ads in magazines and other tourist publications both abroad and locally. Whether tourists arrive in Mendoza by airplane or bus, they will find attractive publicity articles in the airplane or bus magazines.

At the hotel, tourists will find more information about the winery in their room, placed there due to Zuccardi's positive relationship with hotel concierges. Online, tourists can easily find the Zuccardi website, which receives more than 300,000 hits per year. Through the website, potential visitors can find information about visitor center hours, booking a tour and experience, or eating in one of the restaurants. Social media tools such as Facebook, Twitter, and Instagram are also used for promotion. Zuccardi also participates in trade tastings in Argentina and abroad, and invites journalists and bloggers to visit the estate.

RESULTS AND BEST PRACTICE IMPLICATIONS

The case of Zuccardi Winery can be considered a best practice in wine tourism for several reasons. A primary one is the very positive increase in visitors to the winery. For example, in 2001 when the Visitor Center was first opened, they only received around 2,000 visitors per year, but now they average over 40,000 per year (Thach, 2014). Furthermore more than half is international visitors, with many coming from neighboring Brazil.

The increase in visitors also corresponds to an increase in sales revenue both at the winery and in retail shops. Julia reported, "The visitor center and restaurants became profitable after only five years of operation." For such a massive operation, this is quite an achievement.

The Familia Zuccardi is one of the few Argentine wineries that has realized a strong position in both the domestic and international markets. This is because, even in difficult times, the company has always believed the domestic market should be a priority for the whole industry. The first big success was the introduction of Santa Julia, with a strong marketing campaign. The domestic market now accounts for an estimated $30 million USD and 12 million liters of wine. In terms of export, in 2014 they sold around 10 million liters for $35.8 million USD (Caucasia Wine Thinking, 2015). They export wine to 49 countries, with Canada as the first place export country and the US in second place (Thach, 2014).

They have also achieved very positive consumer feedback on social media sites, with *Tripadvisor* awarding them with a Certificate of Excellence. Some recent comments include:

*"**Great winery, open on Sunday no reservations needed**" - We loved this winery. Great tour with expert tour guide who spoke really excellent English. Tasting was nice also, as is their shop where you can buy great wine and olive oil! We bought a bottle of "Tito Zuccardi" (special anniversary blend, around $40), which was the best wine we tasted in Mendoza! (5 of 5 stars Reviewed August 3, 2015)*

*"**Great Wine, GREAT EDUCATION, super great experience**" - Familia Zuccardi is the perfect winery to begin your wine experience in Mendoza. This is a huge winery with 50 years of experience producing a wide variety of wines, and they have numerous educational experiences! (5 of 5 stars Reviewed March 31, 2015)*

Additionally, many of their efforts have also been recognized by both the Argentina and international wine industries. In Argentina, they have won Business Excellence and Wine Business Track Awards. In September 2007, the prestigious magazine, *Decanter*, acknowledged José Alberto and Sebastián Zuccardi as being among the five most influential personalities of Argentine Wine (Zuccardi.com, 2015). Finally, both of their restaurants have won numerous awards for cuisine and service.

FUTURE ISSUES

Despite its success, Zuccardi Winery still faces several issues. A primary one is the economic condition of Argentina. With high inflation rates and a quasi-fixed dollar-peso ratio, all wineries find themselves struggling to keep prices stable for export. Taxation rates have also grown considerable, generating extra pressure for companies.

Keeping abreast of changing consumer needs and understanding the global market is another continuing challenge for Zuccardi Winery. They recognize they need to increase marketing intelligence and link this to their research and development activities. By focusing on innovation to create new products to match consumer needs, they are trying to stay one step ahead of the changing wine market.

Finally, working in a third generation family business is always challenging. Though they are linked by the bonds of family, and hold fast to their four values of quality, innovation, environment, and community, differences in opinion arise frequently. The Zuccardi Family must continue to strive to work together as a team, while recognizing that diversity of opinion can be healthy when it is channeled towards the accomplishment of their goals, values and what is best for the customer.

DISCUSSION QUESTIONS

1. What impact does the economic situation in a country have on wine tourism? Identify pros/cons for Argentina wine tourism.
2. Zuccardi utilizes innovation as part of their competitive strategy. In your opinion, is this a good strategy, and will it work in the long term? How can you measure whether or not innovation pays off?

3. Zuccardi is a third generation family winery. Succession planning research shows that businesses often start to fall apart when the third generation takes over. Do you believe this is true? What can the Zuccardi's do to ensure their winery continues as a profitable family business?

4. Zuccardi winery has adopted progressive environmental protection practices with their sustainable and organic farming. What else can they do to expand in this area? Do you believe this type of focus pays off in the long-term? Discuss why.

Chapter Twelve

Creating a Unique Wine Tourism Experience: The Case of Moorilla Estate in Tasmania, Australia

Marlene Pratt, *Griffith University, Australia*

It was in 1947 that Italian textile merchant Claudio Alcorso founded Moorilla Estate on a small island in the Derwent River, fifteen kilometers north of Hobart, Tasmania. The term "Moorilla" means a rocky place beside the water, and the land was originally a meeting place for Indigenous Tasmanian people, the Parlevars.

Though formerly the island was planted to fruit orchards, Claudio's dream was to install a vineyard and make wine. Finally, in 1958, he trailed 90 riesling cuttings on the site. The first vintage was made in 1962, making Moorilla Estate the second vineyard to be established in Tasmania. Unfortunately, he encountered some financial struggles and eventually sold the estate to the current owner, David Walsh, in 1995.

David had actually grown up in Hobart, Tasmania in a middle class family, but amassed a fortune when he invented a gambling system to bet on horse racing and other sports. With a strong interest in wine and art, David purchased the winery hoping to make it a cultural center for both. However, the first several years were quite challenging as the winery didn't have strong brand recognition, and competition was heating up with many new wineries opening around Hobart. Wine tourism had not yet come to Tasmania, and it was difficult to lure visitors to a location that was not known for quality wine.

This chapter explores how David Walsh was able to transform a struggling unknown winery into a world-class wine tourism destination. Based on David's vision of exploration, he and his team have created a unique collection of art and visitor experiences that have allowed Moorilla

Estate to be named one of the Best Of Wine tourism initiatives in Australia (Tourism Australia, 2013).

OVERVIEW OF WINE IN AUSTRALIA

According to Wine Australia (2014), vines first arrived in Australia with the First Fleet in 1788. Significant vineyards were planted near Parramatta in 1805 by Gregory Blaxland and near Camden in 1820 by William Macarthur. In 1831 James Busby collected over 362 varietals which were finally planted in Sydney and his Hunter Valley property. Ensuing cuttings made their way to various parts of NSW, Victoria, and South Australia.

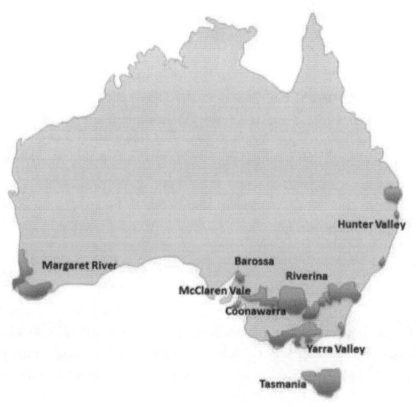

Major Wine Regions of Australia

Many of Australia's old vines can trace their history to the original Busby collection. The Hunter Valley was the first commercial region with Wyndham Estate being established in 1828. By the 1840's viticulture was established by Italians in Riverina, Swiss in Victoria, Dalmatians in Western

Australia, and Lutheran Germans in South Australia, particularly the Barossa and Clare Valleys. According to Wine Australia (2014), commercial viticulture was established in most states by 1850, with the first wine export to the United Kingdom formally recorded in 1854 of 6,291 liters (1,384 gallons).

Within Australia, wine regions are labelled through a Geographical Indication (GI), an official description of an Australian wine zone, region, or sub-region designed to protect the use of the regional name under international law. GI is similar to the Appellation naming system used in Europe, but less restrictive in terms of viticultural and winemaking practices. The only restriction is that wine which carries the GI must include at least 85% fruit from that region (Australian Wine and Brandy Corporation, 2010).

There are currently 65 official GI wine regions located on the southern half of Australia (Wine Australia, 2013). Western Australia has 9 GI regions including Swan District and Margaret River; South Australia has 18 GI regions including Barossa Valley, Clare Valley, and Coonawarra; Queensland has 2 GI regions including Granite Belt; New South Wales has 14 GI regions including Hunter Valley and Mudgee; Victoria has 21 GI including Yarra Valley and Mornington Peninsula; whilst Tasmania currently has only one official GI.

# of Wineries in Australia	2,573
# of Wineries in Tasmania	113

Australia was the 6th largest wine producer in the world in 2014, (OIV, 2015) producing 12Mhl, and the 5th largest wine exporter. South Australia has the largest area of vineyards accounting for 48.0% (71,310 ha) of the national total vineyard area. New South Wales follows with 39,097 ha (26.3% of the total) then Victoria with 25,409 ha (Australian Bureau of Statistics, 2012). There are now over 100 different varieties planted in Australia, however, three varieties of grapes accounted for 60% of all wine grape production in 2012: shiraz (362,217 tons), chardonnay (384,283 tons), and cabernet sauvignon (207,558 tons). A recent interest in other Mediterranean varieties has seen small but increased plantings such as vermentino, barbera, sangiovese, and nero d'avola. Pinot gris and savagnin were the only premium varieties to show growth in vineyard area and increased by 6.6% and 17.9% respectively (Australian Bureau of Statistics, 2012).

Within Australia there are 2,573 companies who commercially sell wine in 2014 (Winebiz, 2014). While the rate of growth is slowing, in the past 12 years there has been an average net gain of 97 wine producers per year. Currently, Victoria has the greatest number of producers with 773, followed by South Australia with 720 and NSW with 484. Tasmania has had a steady increase in the number of wine producers with 113. The wine industry has a large component of small producers, where nearly three quarters (72%) of the companies who provided tonnage figures crush less than 100 tons per annum. There was substantial growth in producers crushing less than 10 tons as 557 companies now represent 22% of all producers. The top four wine companies accounted for about 48% of the national crush in 2013, while the top 20 companies accounted for 87% (Winebiz, 2014). The remaining 2,553 companies then have to compete for the remaining 13% of wine sales.

As wine production has increased, so too has the consumer's consumption and interest in wine. This is evident with Australia's consistent increase in wine consumption, making Australia the 13th consuming wine nation per capita in 2012 (OIV, 2013) up from 19th in 2010. Wine consumption has a strong relationship to wine tourism behavior (Alant & Bruwer, 2004), and is one of the key motivators to engage in wine tourism (Getz, 2000; Hall, 2003).

Wine Tourism in Australia

Within Australia, the image of wine has developed into one of a lifestyle product (Winemakers Federation of Australia, 2002), and correspondingly there is an increasing trend where wine tourism is viewed as a personalized experience where travelers can experience culture, lifestyle, and territory (Hall et al., 2000). This is evident with the increase in wine tourists and the number of wine producers who have a cellar door.

Development of wine tourism as part of the tourism economy has been recognised by the federal and state governments. All states within Australia (except Northern Territory) have developed a separate wine tourism strategy. The Australian wine industry's *Strategy 2025* document stated that wine tourism will be a major source of profits for the wine and tourism industries and a driver of economic, social, and identity development in the regions, which is a key factor to improve the profitability of wineries (Australian Wine online, 2006).

A winery's cellar door or tasting room has become an important conduit for wine sales, for developing and maintaining relationships with clients, and reflecting the image of the winery (Hall & Mitchell, 2008). The number of Australian wineries with a cellar door rose from 1,614 in 2009 to 1,626 in 2015, achieving a total average of 66% (Winetitles, 2015). Victoria has the highest number of cellar doors ($N = 495$) followed by NSW ($N = 356$), and Tasmania has 73 cellar doors (64.6% of all wine producers).

During 2009, there were just fewer than 5 million visitors who visited a winery while travelling in Australia (Tourism Research Australia, 2010). Of these travelers, over 4.1 million were domestic visitors and 660,000 were international visitors. Currently, only 5% of international tourists to Australia specifically visit its wine regions; however, this is increasing at a faster rate than total international visitors. The international winery visitors spent $4.9 billion on trips to Australia, whilst domestic overnight winery visitors spent $2.2 billion on their trips during 2009 (Tourism Research Australia, 2010).

The current environment has made it difficult for smaller independent wineries to sell their wine at profitable levels to export markets, domestic distributors, and retail channels (White & Thompson, 2009). Regional destinations also find it difficult to compete with larger tourism destinations which have capital resources to market and promote their destination. As a result, the majority of the 1,711 wineries with a cellar door are small wineries which are reliant on domestic travelers for not only wine sales, but also the viability of their cellar doors. The importance of domestic tourism is vital to sustain smaller regional wineries.

OVERVIEW OF TASMANIA WINE REGIONS

The first commercial vineyards were planted in Tasmania in 1865; however, the industry collapsed a decade later due to the gold rush on the mainland (Walker, 2014). Interest started again in the late 1970's, and in 1974 licensing laws were released that allowed locally operated vineyards to sell wine directly to the public.

Tasmania has a reputation for quality, cool climate, boutique winemaking that has seen substantial growth over the last ten years, and is predicted to continue to grow. The Tasmanian Premier, Lara Giddings, stated the government plans to quadruple wine production over the next decade through further investment (Cullen, 2013). Tasmania currently has

72 vineyards, 113 wine producers, and 73 cellar doors, with 84% crushing less than 100 tons (Winetitles, 2015).

Yields for 2013 vintage were 20-30% up on the previous season (which produced 7,400 tons) (Wine Australia, 2013b). The state has approximately 1% of the national vineyard area and produces 0.5% of Australia's wine (Dept of State Growth, 2014). The 2013 vintage had 43.5% yield from pinot noir (4950 tons), followed by chardonnay (23.3%), and pinot gris (10.6%).

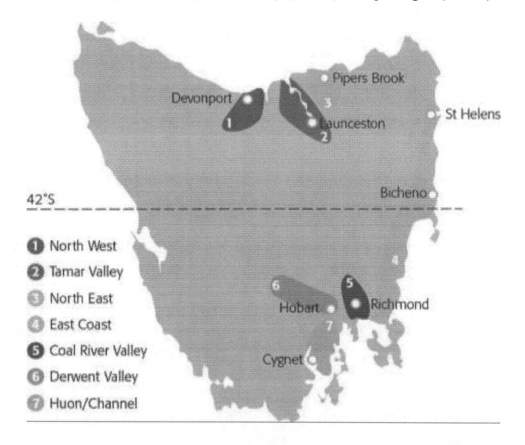

Major Wine Regions of Tasmania

Tasmania is situated off the southern coast of Australia in the cool waters of the Southern Ocean. The climate varies from high humidity in summer to spring frosts in the North to genuine cool climate in the south of Tasmania. Tasmania's climate has been compared to parts of New Zealand and lies at a similar southern latitude to wine producing regions in Bordeaux, Chianti, Oregon, and Champagne (Dept of State Growth, 2014).

Tasmania has a significant advantage with natural resources such as water. An ongoing priority of Tasmania is the completion of major irrigation schemes which will double the amount of irrigated land (Dept of State Growth, 2014). Tasmania's diverse landscape offers diversity in soil, from gravelly basal on a clay and limestone based in Tamar Valley in the north to sandstone and schist in Derwent Valley in the south.

Tasmania is defined as only one official GI but there are 7 geographically distinct regions producing wine. Within the northern part of Tasmania, Tamar River Valley, and Piper's River are located. While in the south of Tasmania is Richmond, Derwent River Valley, Coal River Valley, Huon Valley, and Channel wine regions can be found (Wine Australia, 2013a).

In terms of signature grape varietals, Tasmania is especially well known for pinot noir, chardonnay, and sparkling wines. Tamar Valley and Pipers River have proven to be high quality areas for pinot noir, which is distinctly fragrant and lighter bodied with delicate flavors of red apple and cherry. Tasmania is also renowned for excellent chardonnay, which is described as elegant, complex, and subtle style with high natural acidity (Wine Australia, 2013a). Given the focus on pinot noir and chardonnay, it is not surprising that Tasmania produces delicious sparkling wines made in the Traditional Method used in the classic Champagne region and have developed a world class reputation for their delicate fruit intensity, balance, and finesse. Other grape varieties include riesling, sauvignon blanc, and pinot gris.

The number of winery visitors to Tasmania has also increased as the state has built its reputation for fresh produce, premium food produce, and a profile of gourmet food and wine. Visitors to wineries has increased from 145,914 in 2007 to 207,916 in 2014 (Tourism Tasmania, 2015b).

THE PROBLEM: LACK OF RECOGNITION FOR WINE OR TOURISM

When David Walsh purchased Moorilla Estate in 1995, he acquired a winery and a small restaurant. Unfortunately, the winery did not have a good reputation, because its focus had been on producing high quantities rather than high quality wine. Because of this the winery struggled to achieve success in the market.

To make matters worse, tourism within Tasmania at the time remained low, with wine tourism visitation even lower. However, in 2001, David's vision to create an art center was realized when he opened the Museum of Antiquities at Moorilla. This helped to increase interest in the estate together with tourism interest in Tasmania. According to Tourism Tasmania (2015b), there were 526,336 visitors in 2001, with an increase to 619,015 in 2002. Unfortunately in 2007, the museum was closed for redevelopment.

At the same time that David was working to increase recognition for Moorilla, there was a massive 80% increase in the number of wineries from 1995 to 2008 (Winebiz, 2014). As new wineries became established in Tasmania so too did the competition for visitors. Yet Moorilla Estate continued to perform poorly during this time, and David realized he needed to make some changes. Thus he embarked on a journey to transform Moorilla from an unknown winery to the world-class wine tourism destination that it is today.

THE SOLUTION: DESIGNING A PLACE OF DISCOVERY AND UNIQUE EXPERIENCES

David Walsh's vision for Moorilla was to create a place of discovery and exploration for visitors to engage all of the senses. So over the course of several years, working with his team, he created a destination where tourists could not only stop to taste high quality wine and eat at an award winning restaurant, but they could stay overnight in luxury accommodations, visit a new art museum, participate in a myriad of events, and even taste beer in the estate micro-brewery. In order to accomplish this, he implemented the following actions.

Crafting High Quality Tasmanian Wine

One of the first steps David took was to bring in a new winemaker to upgrade the quality of the wines. Therefore, in 2008 he hired Conor van der Reest as chief winemaker. Conor was originally from Canada, but his winemaking experience in the cool climate regions of Niagara Peninsula and Champagne made him ideal for the job. After Conor's first vintage, he recognized that the existing winery was not sufficient to create the quality wines that were expected, and in 2009 he talked David into building a new wine production facility.

Another decision was made to reduce the production from 500 tons to 150 tons in order to focus only on high quality wines. Since the estate owned two vineyards; one of 1.5 hectares on the winery grounds and a second vineyard of 16. 5 hectares in Tamar Ridge, they were able to focus on small batch production and sell excess grapes. Now their current production is around 9,500 cases.

Today Moorilla Estate has a fairly comprehensive winery portfolio of 20 wines in three ranges of wine:

1) **The Cloth Label series** - The cloth label is their reserve range using labels made of cloth similar to that which the original owner, Claudio Alcorso, used to on his original bottles.

2) **The 'Muse' series** is their mid-range series which focuses on expressing Tasmanian terroir. The Muse labels are black and white with a minimalist design incorporating vines and/or leaves. The earlier Muse labels were choreographed by a Melbourne contemporary artist and consisted of semi-naked images of ballet dancers. These labels were designed to fit with the image and launch of the opening of the museum. These labels were considered too controversial and not accepted into restaurants or their export market.

3) **The 'Praxis' series** is their new world style wine which is soft and fruit driven. The labels of the Praxis series are images of graffiti taken from buildings in different city locations with each new vintage. This further provides an association of Moorilla wines with the museum to create interest in visiting the winery.

Designing and Staffing a State of the Art Cellar Door

The next step was to upgrade the cellar door/tasting room on the property and invest in training and development of employees to greet visitors. The cellar door is situated in the centre of the property in the Ether building. The Ether has a reception area on the ground floor to direct visitors to one of four locations: the cellar door for wine tasting, the event space, the restaurant, or to the hotel if they are checking in for a longer stay.

The tasting room on the first floor has views of the Moorilla vineyard, Mt. Wellington, and the Derwent River. Wine tasting is offered for $10 per person, which is redeemable upon purchase. There are 17 wines available for tasting, and these are shown on a palette format with tasting notes and images of the wine. The tasting is conducted in a circular bar space with seating for approximately 12 people. However, there is space with additional bar tables and chairs a few meters from the bar tasting space. Moorilla uses special glassware that Daniel McMahon, the Cellar Door Manager, states "bring out the floral and spice, and shows the character and integrity of our wine".

Moorilla Visitor Center

Cellar doors are difficult to operate and ensure a profitable investment, so Moorilla feels it is important to make the winery experience engaging and personal. The cellar door staff are trained to ensure they offer professional cellar door service, which includes eye contact, a warm welcome, and friendly conversation to engage visitors. Daniel McMahon believes "it is essential to create a personalized experience." There are four full time employees in the cellar door to ensure a good level of service.

Food and wine matching is also emphasized in the wine tasting to create memorable moments with their wine. David Walsh does not want to sacrifice the experience and make it perceived as a commercial business. As a result, cellar door employees do not focus on wine sales, but on providing a unique experience to their visitors. Alongside the tasting room are books, artwork, and Moorilla merchandise available for purchase.

Mona – Museum of Old and New Art

MONA (Museum of Old and New Art) was opened officially in 2011. It is considered to be Australia's largest private museum located on a wine estate. The entrance to the museum has no signage and is very minimalistic, which is based on David's philosophy of exploration and creating an artistic experience. Three levels of artwork are on display focusing on the topics of sex, death, and evolution. Private artwork owned by David Walsh is also on exhibition, as well as access to his private library. The museum has now become a major tourist attraction for visitors to Hobart.

MONA – Museum at Moorilla Estate with Vineyards

Special Tours of Moorilla Estate

Moorilla offers several different tasting and tour packages for visitors. Providing tours is considered important because many small wineries in Tasmania do not offer this service. The standard tour at Moorilla is a one-hour winery tour and tasting package available Wednesday through Monday at a cost of $15 per person. The guided tour consists of a tour of the vineyard and the gravity-assisted winery. Tasting is conducted during the tour with sparkling wine in the vineyard, another taste in the barrel room, and then a final selection of wines in the tasting room.

Special day-long or multiple day tour packages are also available. These may include transportation from Hobart, meals at the Moorilla restaurant, tours of the art gallery, lodging, and private tastings. Examples include: the Moorilla Wine Meets Mona Art Day Tour, Moorilla's Posh-As Day At Mona, The Moorilla Sleepover At Mona, and The Moorilla Magnum. These tours range in price from $137 to $3600 per person.

The Source Restaurant and Estate Cafés

The *Source* restaurant is a fine dining restaurant which can hold up to 80 guests. The executive chef, Philippe Leban, has a Michelin star-studded resume, and specializes in contemporary, French-inspired cuisine with Tasmanian influences. The menu offers a contemporary style breakfast and lunch, with a degustation for dinner. The wine list includes wines from Moorilla estate, other regions of Australia, and also an international selection.

As part of the development of the winery, a café and bar was designed to showcase the winery. The café has a glass wall which allows visitors to view the barrel room and the winery. The café offers contemporary food and beverages including Moorilla wine.

Within the museum, there is also a café and separate bar, however, these food and beverage outlets are targeted toward the museum patrons. As a result of the demand for bread and pastries, Moorilla has a full time pastry chef, and all bread and pastries are now made in-house. This provides a better quality product in the restaurants and cafes, as well as ensuring consistent quality and delivery.

Luxury Accommodation in Moorilla Pavilions

There are eight luxury pavilions on the property, and each one is named after the artwork on display in the apartment. For example, the 'Sidney' pavilion has Sidney Nolan artwork throughout the apartment. Four of the pavilions were built in 2006, and then, due to popularity with visitors, four more were added in 2011. Accommodation includes breakfast in the Source restaurant, wine tasting and tour, and entry to the museum. Other facilities include an indoor oxygenated swimming pool, gymnasium, sauna, and tennis court for the guests.

Unique Boutique Microbrewery – Moo Brew

The owners of Moorilla also have a boutique microbrewery, Moo Brew Brewery, which was built on the estate in 2000. They expanded the brewery in 2010 by opening a second brewery site ten minutes from the winery in Bridgewater. Due to continued expansion, all brewery operations are now carried out at this secondary site. The cellar door offers Moo Brew beer flights for $10 and a Moo Brew Tour and tasting for $30 per person once a week.

Multiple Events at Moorilla

Moorilla has a range of events and concerts held on the premises. The music events include local musicians as well as national and international acts such as Cat Empire, and Southern Gospel Choir. These events are held on their permanent stage which is located near the Ether building and can comfortably accommodate up to 4,000 patrons on the lawn. Other events include Mona Market days developed by Kirsha Kaechel, David Walsh's partner, which supports local artists and food producers.

Moorilla is also involved in hosting two festivals per year in Hobart together with MONA. The MOFO four-day music festival is held in January and a second festival is held in the middle of winter. The first festival was held in 2012, with a higher than expected attendance of 10,000 people, followed by an astounding increase to 15,000 attendees in 2013.

Moorilla has great array of venue spaces to hold a range of functions, from corporate meetings to special events and weddings. There are two meeting rooms: Eros and Thanatos located in the Ether Building. Each room can cater up to 150 people or can be combined into a large function space. In addition, there are venue spaces within the museum which can hold cocktail receptions or dinner functions. Moorilla recently hosted a special dinner for Tourism Australia as part of their Restaurant Australia campaign. Over 300 VIP guests attended, and were treated to a special meal with world renowned chefs Neil Perry, Peter Gilmore, and Heston Blumenthal.

Transportation to Moorilla – Via Land and Water

Moorilla is fifteen minutes from the city center of Hobart, but to ensure continued and convenient visitation to the museum and winery, there are various transportation options, which have been introduced. The main form of transportation is a catamaran which departs from the marina in Hobart city every hour between 9:30am and 6pm and takes visitors to the island where Moorilla is located. The catamaran has two categories of seating: standard and the posh pit. The posh pit includes access to an exclusive lounge, bar, and private deck where complimentary beverages, including Moorilla wine and Moo Brew beer are served, along with complimentary canapés, pastries, and antipasto platters. There is also a MONA Roma mini bus which operates alternate times to the ferry.

Water Approach to Moorilla Estate on Island in Derwent River

Sales, Marketing and Sustainability Efforts at Moorilla

The team at Moorilla works hard to reach out to potential consumers and build relationships with existing customers. Their goal is to establish long-term relationships based on the unique experiences and high-quality wines that Moorilla provides. In order to accomplish this, they implement the following sales, marketing, and sustainability efforts:

Wine Sales - Wine sales are the responsibility of the winemaker, Conor van der Reest. Currently, 50% of sales occur onsite through hospitality operations including the restaurant, bar, and events. Twenty five percent is sold through the cellar door and the remaining sales through wholesalers in Australia and select restaurants in Tasmania; however, Conor has a goal to increase cellar doors sales. Only one percent is exported, but Moorilla is slowly building a reputation of producing high quality Tasmanian wines crafted in an old world style, which is helping increase export sales.

Marketing and Public Relations - Cellar door manager, Daniel McMahon, also has the role of Marketing and Public Relations Manager. Public relations take an important place in Moorilla's promotion efforts, with extensive coverage in print and online publications. Other marketing activities include participation in wine shows and wine events, and direct mail in the form of newsletters to club members.

Website - Moorilla's website is designed in a simplistic style with minimal information, but enough information to outline what the owner considers to be their core business. David Walsh wants minimal information on the website in order to create the element of surprise for visitors. As a result, a decision was made not to include, for example, a video or virtual tour of the estate on their website. Moorilla has a twitter account that has been managed by the winemaker sharing vintage information with followers. The business also has a well-developed online buying platform (Moorilla.com, 2015), where visitors can purchase Moorilla wines, a wine tasting and tour package, event tickets, and/or museum entry tickets.

Wine Club - Moorilla Estate offers a Cellar Club membership program which is free of charge. The membership includes: Free tastings and winery tours, newsletters featuring new wine releases, seasonal e-bulletins with upcoming events, news, reviews, specials, and tasting notes, and first notice, technical notes, and availability alerts for preferred wines. The club membership also offers cellaring of wine for release.

Wine Buying Program - Another program is the Wine Buying Program which involves sending out a wine parcel each quarter at the beginning of each season. For example, The Muse Series spring parcel cost AUD$245 and includes two bottles each of sparkling rosé, riesling, and pinot noir wines. Each year the winemaker telephones club members to continue relationship building to stay with Moorilla.

Sustainability Efforts - As with many wineries, the owners have a concern for sustainable practices. Kirsha Kaechele has also been a driving force in sustainability efforts of the Derwent River. With assistance from several universities, a strategy was developed to clean the Derwent River from the heavy metals on the bottom of the river.

RESULTS AND BEST PRACTICE IMPLICATIONS

It is well known that wine tourists seldom visit wine regions only for the wine, but expect a total wine tourism experience which includes a combination of quality wine and regional cuisine in an attractive environment. Motivational factors include an inherent element of hedonism, as the tasting of wine involves alcohol and there are links with food, socializing, and relaxation which point to an indulgent activity (Hall et al., 2000).

Peter Schulz, the President of the Winemakers Federation of Australia (WFA), also acknowledges that producing good wine and food and offering it with a smile at cellar doors and restaurants is not enough. Instead, wine and food regions need to offer a compelling reason for people to visit. Potential visitors need to be able to easily identify and choose from a range of tailored options at all price points to suit differing tastes and moods, as many tourism activities compete for tourists custom (Winemakers Federation of Australia, 2011).

This challenge has been met by Moorilla by offering more than a tasting experience in an attractive setting but through the development of a multidimensional tourism experience. More specifically, Moorilla has captured all elements of the wine tourism product. The cellar door experience offers a wide choice of quality wines to taste with high standards of service quality. This is evident through the achievement of a Five Star winery rating from James Halliday (Halliday, 2015), Australia's premiere wine critic. Winery and vineyard tours are included to provide an opportunity for visitors to explore the process of winemaking and meet the winemaker.

A recent customer stated about their experience at Moorilla on James Halliday's Wine Companion website: *"Easily one of the great wine experiences in the world"*.

More importantly, Moorilla estate offers a very unique artistic experience make a lasting impression on visitors which encapsulates a unique "winescape". The winescape includes not only a vineyard and beautiful surrounds of Mt. Wellington and the Derwent River but artworks spread throughout the estate.

Personal development and a sense of inspiration about the wine tourism experience has also been identified by Sparks (2007), which is in keeping with David Walsh philosophy of exploration and the art gallery. Visitors to MONA have increased over the last three years when *Tourism Tasmania* started collecting data, from 210,300 visitors in 2012 to 300,900 in 2014. This is an increase of 43%. MONA is now the second most visited tourist attraction in Hobart (Tourism Tasmania, 2015a).

Cellar door visitation also had a remarkable impact from the opening of the art gallery. Visitation in 2009/2010 prior to the opening of the art gallery was 10,800, which increased to 18,780 by 2014/2015. This has resulted in an increase of 74% of cellar door visitors.

Moorilla is very active in hosting events and festivals on their estate and also within the city of Hobart, which is increasing in attendance to a record of 40,000 attendees in 2015 (ABC News, 2015) and over 130,000 people attending each of the two Dark Mofo festivals.

Finally, Moorilla is part of the 'Ultimate Winery Experiences Australia' launched in 2013 that is a hand-selected collection of Australia's premium wineries "offering quality winery experiences based around world class wines, warm and knowledgeable hospitality and culinary excellence" (Cope, 2015). Each of these attributes has been developed and integrated into the unique wine tourism experience that can be found at Moorilla Estate in Tasmania.

FUTURE ISSUES

Though Moorilla has been very successful at attracting large numbers of visitors and local customers who return again and again, there are still a few issues with which the estate grapples. One is actually the burden of success and growth, in that there is often not enough parking for visitors.

Additionally, David's desire for lack of signage to encourage exploration is sometimes challenging for some visitors.

As wine production increases, so too does Moorilla's need to increase sales via their various channels: cellar door, domestic retail, and export markets. This requires strong marketing and branding skills, as well as a need to stay close to the customer and keep abreast of changing consumer desires.

Finally, as with any privately owned winery, the owner has a strong influence on the direction of the business, which should ideally, will align with the desires of the winemaker and the rest of the management team. The vision and mission of the business needs to be cohesive, particularly with the multidimensional experience that Moorilla Estate and MONA offer. Some question whether it is the wines of Moorilla or MONA that tourists are coming to visit. Should Moorilla Estate create a separate identity from MONA, or are they too intermeshed to divide…and does it make sense to do so from a tourism perspective?

DISCUSSION QUESTIONS

1. Discuss the wine tourism product Moorilla is providing its visitors. How can it be improved?
2. Identify some problems Moorilla's cellar door may have in attracting visitors?
3. Discuss the positive and negative impact of the owner's philosophy on the wine tourism operations of Moorilla.
4. Management would like to increase the amount of time visitors spend on the estate and the average spend. How can they do this?
5. Do you feel that the image of winery needs to compliment or reflect the image of the museum?
6. What future recommendations would you give to the owner, the winemaker and the cellar door manager to improve their business?

Chapter Thirteen

Wine and Kids: Making Wine Tourism Work for Families in Beaujolais at Hameau Duboeuf

Joanna Fountain & Laurence Cogan-Marie

Lincoln University, NZ & School of Wine & Spirits Business, ESC Dijon

Picture the scene: you are enjoying a relaxing day visiting a winery, doing some wine tastings, having a meal, and taking pleasure in the surrounding countryside. Suddenly the tranquility is shattered by the sound of children fighting over whose turn it is to use the iPad. You turn to see the harassed parents quickly setting down their wine tasting glasses, before beating a hasty retreat with their children.

Or another scene: you are a busy professional spending a family holiday in the countryside, off to another amusement park, but your heart is not in it. Staring wistfully out the window you watch the vineyards flash by and signs enticing you to stop and taste at a local winery. You sigh and wonder if the days of wine tourism will have to wait until your children grow up.

There is little doubt that wine tourism is not primarily viewed as an attraction for family visitors. Indeed, family groups make up a small proportion of visitors to wineries in most regions of the world. This is made clear with a quick overview of wine tourism promotion, where one finds very few images that promote wine experiences that include families and children.

There are many reasons for this exclusion of families from wine tourism activities, including a general reluctance to view attractions serving or promoting alcohol in any form as an attractive or appropriate activity for children. There is a sense, also, that what is being showcased at wine tourism attractions is of little interest to children–whether that be the wine itself, or the relatively complicated scientific process of grape-growing and winemaking. However, it doesn't have to be this way.

This chapter presents a case study that illustrates how Hameau Duboeuf, a pioneering wine tourism attraction in Beaujolais, France, is defying these perceptions. Called the "First Theme Park of Viticulture and Wine," Hameau Duboeuf provides a wine experience that appeals to both adults and children, enabling families to engage in a fun interactive family time together.

OVERVIEW OF WINE TOURISM IN FRANCE AND BEAUJOLAIS

It is a well-known fact that more international tourists visit France than any other country each year (UNWTO, 2012). France is a very dominant player in the global wine industry in terms of producing some of the most famous wine labels in the world, as well as some of the highest production rates. Indeed in 2014, France was the largest global wine producing country with 46,151 thousand hectoliters, ahead of Italy at 44,424 and Spain at 37,000 (OIV, 2014). It also boasts more than 85,000 wine producers (FranceAgriMer, 2013).

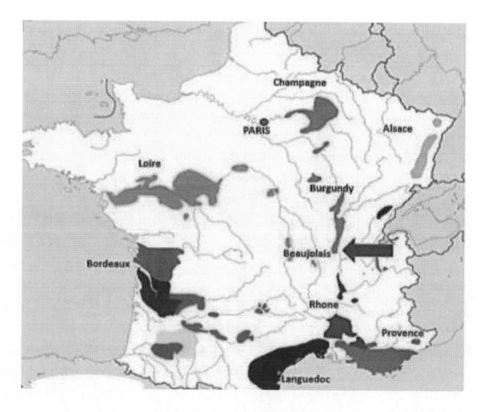

Location of Beaujolais in France

Despite the widespread recognition of many of the famous wine regions of France, this does not necessarily translate into a well-developed wine tourism industry, or significant participation in wine tourist activities in some of these regions (Frochot, 2000; van Westering & Niel, 2004). Overall, however, it is estimated that ten million people participated in French wine tourism in 2010 by visiting a wine cave or chateau, with 40 percent of these being international visitors (Atout France, 2010).

OVERVIEW OF BEAUJOLAIS

Beaujolais, which is administratively still considered to be the southern part of Burgundy (Robinson, 2006), is located about 45 kilometers north of Lyons. It is world renowned for its Beaujolais Nouveau wine, which is released each year on November 17 to celebrate the vintage of the gamay grape in Beaujolais. The young wine, aged only a few months, is fresh, fruity, and not complicated. Its release gives permission for wine lovers around the world to buy an inexpensive bottle and throw a Nouveau Beaujolais party. Originally considered a great marketing concept because of the huge increase it caused in global wine sales, now the downside is that many consumers are not aware of the higher quality and more complex Cru wines of Beaujolais. In general, outside of France, only wine connoisseurs are aware of the charms of the ten Crus of Beaujolais: Chiroubles, Fleurie, Chénas, Morgon, Moulin-à-Vent, Juliénas, Brouilly, Côte de Brouilly, Saint-Amour, and Régnié.

Wine Tourism in Beaujolais

When it comes to wine tourism, Beaujolais mirrors the situation in much of the rest of France. There are a number of wine tourism attractions; including wine caves and chateaux open to the public. There are many winegrowers and wine merchants, with the most recent printing of the Beaujolais Wine Guide Book (2014a) including over 160 wine estates, wine co-operatives and wine growers.

These wine attractions are complemented by the region's fine natural attributes and rich gastronomic and cultural heritage, which can be discovered along the Beaujolais wine route. The route takes the visitor through picturesque hilly countryside and into quaint, historic villages. As in much of France, the region is known for its fine food and the quality of the local produce, and there are a number of fairytale chateaux to visit. The visitor experience of these attractions is enhanced by a series of audio guides, available in French and English, which can be downloaded from the Inter Beaujolais website, a wine marketing organization which manages the promotion of wines and tourism of the region (Inter Beaujolais, 2014b).

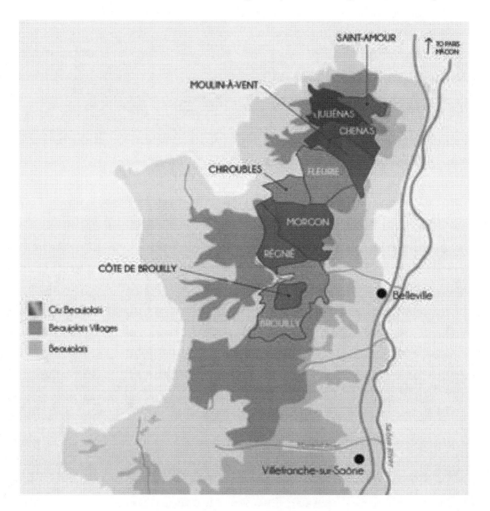

Map of Beaujolais Wine Region

As a wine tourism destination, Beaujolais has many factors in its favor. As well as the attractions outlined above, the region is within easy reach of an international airport located in Lyon, 50 kilometers away, and AutoRoute A6 carries 70 million vehicles each year through the region (Inter Beaujolais, 2014b). The landscape is charming, filled with hectares of vineyards and gently sloping hillsides, nestled amongst which are chateaux and picturesque villages. Compared to other regions of France, many of the chateaux are open to the public and visitors are warmly welcomed.

There are a number of challenges also. First, the region lacks a clear strategy for growing wine tourism in the region, and there is generally a lack of coordination between the various local administrative bodies (Cogan-Marie & Charters, 2014). There is also a sense that the region suffers from being 'little brother' to the much larger and more well-recognized (and regarded) wine regions of Burgundy to the north, and the Rhône Valley, to the south. For example, some experts have argued that the recent fall from favor of Beaujolais Nouveau, and its reputation as a cheap, light-bodied and unsophisticated wine, may limit Beaujolais's attraction as a wine tourism destination (Cogan-Marie & Charters, 2014).

| # of Wineries in France | 85,000 |
| # of Wineries in Beaujolais | 160 |

There is also a lack of knowledge regarding the wine tourism market in the region, and very little information available on the numbers or profile of wine tourists to Beaujolais. However, a study conducted by the Interprofession du Beaujolais (2009, 2010) revealed that wine is the primary reason for visiting the region for half of the Beaujolais tourists, who are, on average, middle-aged (45 years of age) with 65% coming from regions outside of France.

There is no information about the proportion of family visitors to the region, however it is telling that when the filter of 'children' is selected on the interactive wine tourism map presented by Inter Beaujolais, the selection of attractions available to the tourist drops from more than fifty to four. One of these attractions is Hameau Duboeuf, which not only accepts children, but welcomes them with open arms and no entry fee.

THE PROBLEM: MANY WINE REGIONS DO NOT ENCOURAGE OR SUPPORT VISITORS WITH CHILDREN

Recently, a representative from a Burgundy tourism organization was asked, "If one aspect of wine tourism could change in Burgundy, in what area would you like to see improvements?"

Her response was telling: "The region needs to become more welcoming for families. I think we really lose part of wine tourism because we don't know how to deal with families."

This very honest assessment of the Burgundy region's ability to provide tourism attractions that will appeal to all visiting family members is revealing, but it is not unique to Burgundy, or France. Family tourism is a significant market for the tourism industry and yet until recently there has been limited research on the experience of families visiting tourist attractions (Schänzel, Yeoman, & Backer, 2012).

If the attention paid to family tourism is limited, academic research on wine tourism for families is virtually non-existent (Thach & Olsen, 2005). Similarly, there has been very little written in the popular media about the experiences of tourists with children at wine tourism attractions, although wineries and wine tourism attractions will sometimes advertise their appeal to families (Haurant, 2012; Thach & Olsen, 2005).

There are a number of reasons for the lack of attention in the wine tourism literature. First, family groups generally make up a small proportion of visitors to wineries. In Burgundy, a region with considerable wine tourism, only a quarter of all visitors are families travelling with children (Atout France, (2010). For the remainder of France the percentage is somewhat higher, 34% of winery visitors coming as a family (Atout France, 2010). By comparison, families make close to 40% of visits to most other types of tourism in France, such as heritage tourism. This is reflected in parts of the New World also. For example, in New Zealand, only 10 percent of international tourists who visit a winery are travelling in a family group (Tourism New Zealand, 2014).

Given their small market share, it is not surprising that the needs and experience of families are not significantly acknowledged in the literature. Therefore, according to Atout France (2010), "the development of policies for families and children visitors is less of a priority than in other tourism sectors (p. 15)." It is unclear, however, whether they are underrepresented in wine tourism statistics due to a lack of interest or because of the limited facilities available to meet their needs.

For some wineries, the failure to provide for families may be a deliberate marketing strategy, because appealing to families may detract from the brand image of the winery, or the visitor experience of other key market segments (Thach & Olsen, 2005). It may be that the lack of attention paid to this topic is due to the implicit assumption, at least in an English-speaking context, that alcohol and children don't mix. Haurant (2012) explains in her review of family wine tourism experiences: 'wine might be quite civilized, but it is still booze'. She describes her trepidation in taking her three young children to visit wineries in Bordeaux, explaining that she is concerned that she and her husband would be 'condemned as irresponsible parents if we took up the vineyards' tempting offer (to visit with young children)'.

The lack of participation by family groups in wine tourism is a problem for the industry, because the family, including children, represents one of the largest markets for the leisure and tourism industry (Carr, 2006; Obrador, 2012). Extensive research over the past two decades has demonstrated that family holidays, family leisure, and family outings have positive contributions to healthy family relationships (Lee et al, 2008; McCabe et al, 2010; Petrick & Durko, 2013). If wine tourism is to grow, it cannot ignore the needs of this market.

So what is important for families when they participate in tourism activities? One of the very important measures of success of a family tourist attraction is the opportunity it presents for meaningful interaction between family members (Lehto et al, 2009; Shaw, 1997). Spending time as a family helps to strengthen family bonds by creating shared memories, and enhances communication between family members. In addition, attractions that offer opportunities for shared learning provide parents with resources through which to pass on knowledge and skills to the next generation, further strengthening relationships (Erikson, 1950). Therefore it is imperative that if a wine tourism attraction aims to appeal to families, it should provide opportunities for this shared interaction to take place.

It is acknowledged that meeting the needs and interests of all family members at a wine tourism attraction can be challenging, and the reality of 'family time' is often stressful due to competing and conflicting interests (Carr, 2011; Backer & Schänzel 2012). For example, while children primarily seek fun at attractions (Schänzel & Smith, 2014), their parents might be seeking more restful, or educative experiences (Fountain et al, forthcoming).

There is no doubt that some wine tourism attractions do make an effort to cater to families. For older children a winery or other wine attraction might offer a quiz sheet, games, or a treasure hunt to provide some level of education and engagement with the wine experience. For younger children, however, most often what is provided is a source of distraction; coloring pencils and crayons, snacks and drinks, or play equipment to keep the kids quiet while mom and dad enjoy a tasting or a tour. In this situation, there is often limited engagement with other family members and the children typically remain quite peripheral to the wine tourism experience. However, to ensure a truly successful family tourism experience it is necessary to go beyond the standard check-list of kid-friendly activities.

THE SOLUTION: CREATING A WORLD-CLASS FAMILY WINE TOURISM EXPERIENCE LIKE HAMLET DUBOEUF

So what is the solution? Is it possible to provide all family members with a memorable *wine* tourism experience? Georges Duboeuf believes that it is. In the small, unassuming village of Romanèche-Thorins, in 1993, Georges started building what has been described as the 'French Disneyland for Wines' (Allen, 2014). Dubbed the 'King of Beaujolais' (Orlin, 2014), Georges Duboeuf has been working in the region for more than 50 years and is one of the main champions of Beaujolais Nouveau.

Duboeuf's role in global wine tourism cannot be underestimated. At a time when wine tourism was in its infancy worldwide, and largely shunned by French winemakers, Duboeuf set about establishing a wine tourism attraction that he hoped would appeal to visitors of all ages and all levels of wine knowledge.

Inspired by a trip to Disney World, in Orlando, Florida, Duboeuf envisioned a place in Beaujolais where an entire family could come and enjoy the day. What impressed Duboeuf about Disney World was the theme park's ability to provide entertainment to large numbers of people, while simultaneously offering educational experiences to visitors of differing abilities and ages. His son, Franck, described his endeavors in 2012:

> *"My father put his vision out there and managed to breathe life into it. He gave reality to something which did not yet really exist: wine tourism" (Hameau Duboeuf, 2013, p. 4).*

The Wine Tourism Experience at Hamlet Duboeuf

Hameau Duboeuf is housed in a series of buildings that today covers more than 30,000 square feet; triple its size on opening (Prescott, n.d.). The centerpiece of this complex is an extensive 'wine museum' or Hameau du Vin, although this phrase barely does justice to the range of exhibits and experiences offered here.

On arrival at the complex, the visitor encounters the entrance hall and ticketing area, modeled on the Paris railway station Gare de Lyon. From here the tourist enters the exhibition halls. There are all the traditional elements expected in a wine museum, including presentations covering two thousand years of the history of vineyards and wines. On display are more than 3000 items from Duboeuf's own personal collection of wine artifacts and tools, including an eighteenth century screw press. There are a series of other rooms presenting the elements involved in the process of winemaking, as well as barrels, bottles, corks, and an exhibition area decorated as a Beaujolais bistro.

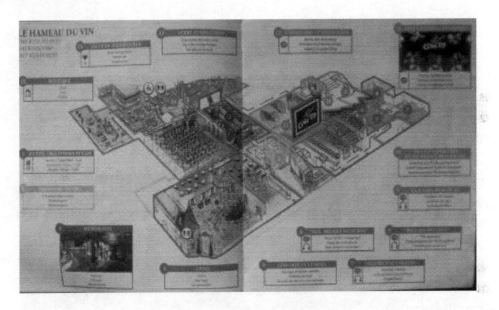

Map of Hamlet Duboeuf

Alongside these static displays are a variety of experiences that engage the senses. There is a sensory experience area where miniature barrels contain the aromas of wine, an automated puppet theatre, a 3D movie experience, and a simulator cinema experience. The visit to the wine museum draws to an end with a wine tasting hosted by a sommelier in a beautifully appointed hall, before the visitor is directed through the gift shop and towards the onsite café.

Beyond the wine museum there are three other sections of the attractions. The first, the Gare du Vin (The Station of Wine) is housed in the Romanèche-Thorins train station that Georges Duboeuf bought when he established the Hamlet. Here Duboeuf is able to cater to his love of wine transportation history, with a range of exhibits harking back to the days when most wine was transported by rail. There are two model trains, each speeding through a panorama of the Beaujolais countryside.

Another section, Le Centre de Vinification (the Winery) gives visitors a chance to view the winemaking process in more detail. The third section is Le Jardin en Beaujolais (Garden in Beaujolais), which surrounds the winery, and incorporates a sensory garden where the visitor can discover the scents and aromas found in wines. In addition there is a rose garden, vegetable garden, and woodland area. For the younger visitor, these gardens also have a mini-golf course, a range of giant games and a maze.

Museum Exhibit at Hamlet Duboeuf

Visitors can ride a small train between the various buildings. All of these activities and attractions mean that it is quite possible for a family to spend five hours or more exploring the site. During school holidays the attraction also holds special activities, specifically for children, such as Christmas displays in December and holiday afternoon teas and games at other times of the year.

From the beginning, Duboeuf's desire was to create a place of learning, fun, and entertainment that would appeal to everyone from wine connoisseurs to preschool children. At the opening of one of the more recent attractions to the Hamlet, the Cine Up, Franck Duboeuf restated his father's vision:

"Our goal is clear: we want our visitors to enjoy meeting us and to be revitalized. We have introduced sensation and liveliness. We are maintaining our aim of teaching through playfulness." (Hameau Duboeuf, 2013, p.4).

RESULTS AND BEST PRACTICE IMPLICATIONS

Since its inception Hameau Duboeuf has been a success, and the continuing efforts made to keep the attraction 'fresh' with new activities and displays, particularly aimed at children and repeat visitors. Because of this, the Hameau attracts more than 80,000 visitors per year (Hameau Duboeuf, *pers. com.*). Today, 75% of visitors are French, the majority of these living within three hours of the attraction. Another significant market is German holidaymakers, particularly those travelling on barges through the region. To support international travelers, the website, brochures and guide maps are available in German, English and French. Not surprisingly families – children with parents and/or grandparents – make up an important target market for this attraction (Franck Duboeuf, *pers. com.*).

The following section outlines some of the ways in which Hameau Duboeuf has created a memorable family wine tourist attraction by offering 1) a wide range of experiences; 2) the engagement of multiple senses; 3) opportunities for family time; and 4) clever pacing and sequencing of the activities.

#1 - Wide Range of Experiences

A primary reason for the success of Hameau Duboeuf is the focus on providing a memorable tourist experience to all visitors, regardless of age. An experiential perspective on wine tourism activities has become a focus of wine tourism researchers over the past decade (e.g. Bruwer & Alant, 2009; Charters et al, 2009; Pikkemaat et al, 2009; Quadri-Felitti & Fiore, 2012, 2013).

Child Exploring Hamlet Activities

Many of these analyses draw on Pine and Gilmore's (1999) experience economy framework. This model delineates four realms of consumer experience, emerging from the intersection of two dimensions – firstly, the degree of customer participation in creating the experience (active to passive) and secondly, the degree of absorption or immersion in the experience, with absorption involving the engagement of the consumer's mind, while immersion results from being physically (or virtually) enveloped by the attraction. Pine and Gilmore (1999) assert that the most successful experiences are those in which all 4 realms co-exist as possibilities for consumers:

- educational experience (active and absorbing)
- entertainment experience (passive and absorbing)
- aesthetic experience (passive and immersive)
- escapist experience (active and immersive)

It is the argument here that in the Hameau du Vin all of these experiences are available. As might be expected of a museum the educational experiences might be the most obvious, via the many artifacts and displays of the tools and processes of viticulture and winemaking.

There are many entertaining experiences also, particularly through the movies and other animated attractions. For example, the four seasons of the vine are described in an electronic theatre, whereby the animated figure of a winemaker tells the story of the vineyard throughout the year. He is supported in this story by an animated vine, with whom he shares a conversation, and with animated townsfolk, who appear illuminated from behind a translucent screen at various points of the story. A three-dimensional film tells the story of the visit of a mysterious young woman to a vineyard to participate in the harvest and release of Beaujolais Nouveau (who the audience knows to be the Roman Goddess Ceres, goddess of fertility and harvest, in modern form) proves an entertaining experience for young and old alike. This film, involving 3-D glasses, also provides a more esthetic experience also, pulling the audience into the action much more than a two dimensional film would do.

Aesthetic experiences are possible also in the large wine barrel room, where the immediate and familiar smell associated with these places transports the visitor to thoughts of other wineries visited. Finally, the most recent addition to this 'museum' the Cine Up, provides the visitor with a truly escapist experience. Entering a pod, the visitor takes a journey up into the clouds with a couple of cartoon bees, who take the visitor on a flight across the fields and hills of the Beaujolais and Mâcon regions in search of pollen, getting into mischief on the way. This is quite literally a moving experience, as the pods tip and move to mimic the bee's flight, and a flight close over the Saône River results in a fine mist being felt by the visitor.

#2 - Engagement of Multiple Senses

It is generally recognized that engaging all the senses is an important way to enhance the engagement of visitors at tourist attractions (Moscardo,

1999). Hameau Duboeuf provides many opportunities to engage those senses often neglected at museums – touch, taste, and smell. There are parts of the museum which are 'hands on', where visitors can explore exhibits in various ways, while also engaging other senses. For example, the sensory exhibit mentioned above allows visitors to open barrels and smell the aromas of wine. The emphasis on the different senses is particularly highlighted for children, with a special quiz attached to the sensory barrels for children.

There is also a 'Touching Room' where they feel around in a dark box to identify a word spelled out in raised woodblocks, and a 'Listening Room' in which children listen to sounds through headphones and record the sounds they hear. Of course, the end of the visit includes a tasting room, where all the senses are engaged as the visitor tastes and smells a selection of the wines, or grape juice, while listening to the description of the wines from a sommelier.

Child Drinking Grape Juice from Wine Glass with Her Family

#3 - Opportunities for Family Time

While these first two themes are arguably evidence of a good attraction in general, the final two themes relate specifically to the success of this attraction as a *family* wine tourism attraction. As stated above, the goal of a family tourist attraction should be to provide the opportunity for all family members to enjoy the experience together, by catering to their different needs.

Hameau Duboeuf uses the traditional technique of including a quiz for children, who have to find the answers to a range of questions using their different senses as they move through the exhibition space. However there are other, more original, ways in which experiences are created that are enjoyable for both children and their caregivers. For example, the Cine Up experience, while primarily aimed at children, provides their parents with a lovely overview of the scenery of the region. Similarly, Noah the first winegrower provides information about winemaking and grape growing of interest to the less knowledgeable adults, but in a way that keeps the children entertained. Finally, the wine tasting area provides grape juice for the children at the counter, and in wine glasses alongside their parents, ensuring that they remain engaged in this very important part of the experience.

4 - Clever Pacing and Sequencing of Activities

It is a known fact amongst parents that sustaining a child's attention at an attraction for long periods of time can be challenging. This is particularly the case for younger children, who tend to have relatively short concentration spans. This is an area in which Hameau Duboeuf excels, as they have managed the pace and intensity of the experience in such a way that even young children remain engaged throughout. For example, the attraction begins with a room displaying historical artifacts of grape growing and winemaking. While of great interest to many adults, the largely static displays are arguably less interesting to children, even with the children's quiz, which starts in this area. Having the museum displays first means children are exposed to it before they get tired and bored.

These rooms are followed quite soon after by the animated puppet theatre. The pacing continues like this, with an activity that is highly engaging for children, proceeded and followed by attractions that have potentially less interest for children. In this way, it is quite possible for a family to spend a couple of hours in this part of the attraction with no one complaining.... an important consideration for successful family outings.

Repeat Wine Tourists and Feedback

Perhaps the best evidence of success of this venture is the high number of visitors to the site, including repeat visitors from surrounding areas. For example, on the occasion of one author's visit in November, a range of giant games were at the disposal of children, who were spending an afternoon playing while their parents or grandparents watched on. After their busy play, the children were treated to a massive 'goûter' (or afternoon tea) with sweets, cakes, and drinks; all this for the price of 10 Euro.

Feedback from Trip Advisor also highlights the success of this attraction for children – in fact there are no negative reviews from people travelling with children. The pleasure of the attraction often seems to come as a pleasant surprise:

Ended up here due to weather not allowing us to sit by swimming pool. Wow, what can I say, entrance fee very good and it gives you access to 3 museums (which you don't have to visit all on the same day). The first museum is huge and takes you through the making of wine through the ages before seeing how the vineyards are kept and when the grapes are harvested. A brilliant interactive cinema experience, which the children enjoyed, and then moving on to seeing how bottles and corks are made before finishing off with a free wine tasting experience. (August 2014, UK).

Some reviewers specifically mention the fact that their children didn't get bored or complain as a highlight of their visit:

Not much to dislike here if you have kids and like wine. Kids were interested to some degree but then when they got a quiz as they went around and a few short movies they didn't complain a bit ... the crazy golf was good and the giant chess was a hit also. (July, 2014; UK)

A place where the whole family spends a great time. An afternoon visit where, for once, our children, 7 and 3 years, were not bored. You learn a lot on the wine estate throughout the visit. A must do (Jan 2015, France; trans.)

A number of reviewers mention how the attractions appeal to both adults and children, with comments such as 'wine history & entertainment for large & small' (Sept. 2013), 'Great for children and adults' (Dec. 2013), and 'Brilliant day out for all the family' (July 2014, UK) being common. For example, a review under the heading 'ideal for small and large' goes on to state:

> *Great time shared between large and small. Everything is done to learn while having fun. We had 6 adults and 5 children (15 years to 9 years) and each has found an interest. Bravo for this site. The cost is not excessive and the restaurant is quite affordable and it's good!! (Oct 2014, France; trans.)*

Even many of those travelling without children noted the appeal for children:

> *Thoroughly enjoyed our visit to this brilliantly designed attraction. It was interesting, informative and along with a delicious lunch in between it was an excellent way to while away 5 happy hours. Would definitely recommend to all who are interested in wines and also for families as well. They have some neat ideas to keep the kiddies amused along the way as well. (October, 2013, UK)*

> *Even though fairly well educated about wine and grape growing, I found the exhibits well done and extensive. I could have done without the bumble bee ride even though I did enjoy the view, but kids would get a kick, and travelling in wine country with children can be difficult to keep them entertained.... The little train ride out to the vineyards and gardens was fun, the views and gardens were lovely, and I'll bet younger visitors would enjoy the mini golf game and huge chess board. (May, 2014, US).*

As the above comment suggests, it is clear that some of those without children were perhaps less enamored with some of the more 'fun' attractions, and for some the child-friendly activities reduced the pleasure of their visit:

> *When I looked into the museum/theme park on the net, I was worried that it might cater a bit too much to kids. I found out that it does....*

However, the wine is good. The museum part of the place is very informative. The sommelier is generous and kind. If you are considering taking a day trip from Lyon to see a bit of the countryside, this is a good choice. ... If you have kids ... this is might even be an excellent choice (July 2014, Denmark)

My wife and I stopped in to do this.... The setting is nice and the place showed a lot of potential. The first room on wine history is excellent, as is the one on the horticultural aspects of growing grapes and the one on barrel making. But the multi-sensory presentations are tacky, geared mostly for children. ....The bumble-bee interactive movie leaves you wondering "what's the point?" and just wastes 15 minutes of your day. (July 2014, US)

Overall it seems that Hameau Duboeuf is achieving what Georges Duboeuf set out to achieve – presenting the history and wine of the Beaujolais region in a way which is both educational and fun, for young and old alike. While elements of the experience offered at this attraction may be found elsewhere – the quiz and sensory games, the 'wine' tasting for children – there are few wine attractions that package a wine experience so clearly for families.

FUTURE ISSUES

The ongoing efforts to modernize and extend the activities for children, and the frequent addition of events and programs to attract repeat visitors, means that the future for this attraction seems positive. There is some evidence, however, that the focus on the needs of families and children may be detracting from the experience of some older visitors. One reviewer suggests that you'd get a more interesting and authentic experience by visiting local winemakers (Sept. 2014, France (*trans.*), and it is true that this attraction would not suit every visitor. However, for businesses wanting to not only attract the family wine tourist but delight them, there is much to be learned from Hameau Duboeuf.

There remain some ongoing issues, as outlined above, about the association of alcohol-related attractions and family entertainment. There has been a backlash recently against the association of alcohol with sporting or culturing activities at least in part due to the effect on children (e.g. Cody

& Jackson, 2014; Kelly et al., 2013), and further development of wine tourism for children could face similar challenges if some parents feel uncomfortable about exposing their children to alcohol in this way.

Alternatively, the Hameau Duboeuf may face challenges from other wineries as they seek to capture this same market. For example, there are a range of other wine-related businesses in the Burgundy region which are offering visitor experiences for children, including the Imaginarium and Cassissium, both in Nuits-St-Georges, and Veuve Ambal in Beaune.

DISCUSSION QUESTIONS

1. What arguments could you make to counter the argument that wine tourism is not a suitable attraction for families and children?
2. Thinking about a wine tourist attraction you are familiar with (a winery, wine museum, etc), how could this attraction apply some of the techniques used at Hameau Duboeuf to appeal more to families?
3. Do you think the existence of more family-friendly wine attractions like Hameau Duboeuf will increase the proportion of family groups visiting wineries or wine tourism attractions? What are the limiting factors?

Chapter Fourteen

Reclaiming a Lost Heritage
at Gerovassiliou Winery (Ktima), Greece

Caroline Ritchie & Kostas Rotsios
Cardiff Metropolitan University, UK & Perrotis College, Thessaloniki, Greece

In the 1970s, a new style of wine burst upon the world stage. It was driven by Australian wine producers creating a new fruit driven style of wine in contrast to the often leaner styles of the Old (wine) World, particularly France. Entrepreneurs from across the Old World looked at this phenomenon and wanted to be part of it. In Greece, it was Yiannis Carras, a wealthy shipping magnate, who initially rose to the challenge. Following the advice of Émile Peynaud of Bordeaux he built the first Modern Greek boutique winery, Château Carras in Sithonia in Macedonia. Unfortunately, the first vintage was a disaster because the winemaker made many mistakes, and most of the wine had to be poured away. Yiannia Carras then hired one of Émile Peyaud's oenology students, Vangelis Gerovassiliou, as his new winemaker and the wines of Château Carras became an international legend (Stevenson, 1997).

However, Vangelis had another dream, beyond that of producing internationally renowned wines from international grapes and marketed using a replica of the French Appellation d'Origine Contrôlée system. He wanted to rescue high quality indigenous Greek varieties from obscurity and to reawaken his fellow Greeks, as well as the rest of the world, to their lost oenological history. He believed that the Greek wine industry "*had lost their knowledge of their wine cultural heritage*". Therefore, in 1981 Vangelis began to renovate his family vineyard in order to develop a winery, which would become a wine-tourism destination attracting and educating a range of national and international visitors.

OVERVIEW OF THE GREEK WINE INDUSTRY: PAST TO PRESENT

As Johnson and Robinson (2013) point out, the modern winescape of Greece is potentially one of the most exciting in the world. It can trace its lineage back to Ancient Greece, which was the cradle of the wine world as we understand it, and having retained many excellent indigenous varieties, is beginning once again to play a part on the international wine stage. The cultivation of the vine for wine making originated in the mountainous regions between the Black and Caspian Sea around 6000 BC (Unwin, 1996). From here, knowledge of viticulture and wine consumption moved south to Ancient Egypt, where it was an elite product, and westwards to Greece via Minoan Crete. By about 2000 BC, wine was being consumed throughout Greece by all levels of society (Estreicher, 2006). As Mycenae, on mainland Greece, emerged as the regional power so the vine and wine increased its significance to Greek society. By the 7th century BC the Ancient Greeks considered anyone who did not habitually drink wine and/or preferred beer, to be inferior and or foreigners.

Major Wine Regions of Greece

Wine was incredibly important to Ancient Greek society as an accompaniment to food, as a trade item and as a social facilitator in symposiums, festivals, and other cultural events (Estreicher, 2006). Greek colonization of the eastern Mediterranean spread the cultivation of the vine to southern Italy, Sicily, and southern France (Johnson, 2006). Its importance is reflected in and on the artifacts and buildings, which survive throughout Europe and parts of Asia. The decline of the Greek empire and the domination of Greece for many centuries by the Muslim Ottoman Empire saw the cultivation, quality, and social importance of wine decline dramatically in Greece (although it continued to flourish in more western countries such as Italy, France, and Spain). By the 1970s and 80s many people's view of Greek wine was colored by the poor quality restinas and oxidized wines offered in both restaurants in cities and local tavernas in holiday resorts.

However, beginning in the mid-1960s, and fully established by the mid-1980, interest and investment in quality Greek wine began to return. By 1991, the five large companies, plus a few co-operatives, who had been supplying almost all of the wine consumed in Greece, had been significantly challenged by these new independent operators who had taken 15% of the market (Jordan, 2000). Many of these "newcomers" such as Vangelis Gerovassiliou had been trained in modern methods outside Greece; some in Bordeaux, others at the universities of Adelaide and Davis in Australia and California. Hugh Johnson and Jancis Robinsons' (2013) "modern revolution" had begun with the enormous challenge of reclaiming the high quality image once held by the wines of Ancient Greece.

Greece is the 7th or 8th largest wine producer in Europe (FAS, 2014: Timothy and Boyd, 2015). It has over 600 commercial wineries, the majority of which are small and family owned. Although declining, vines are grown throughout the country (ELSTAT, 2014), which is split into 28 wine regions including the islands. The most important of these appellations are Nemea in the south of Greece in the Peloponnese region and Náoussa to the north in the Macedonia region.

All styles of wine are produced in Greece, from the sweet Muscats of Sámos to the powerful reds of Macedonia. Some of the most famous Greek grape varieties (Isle, 2009) include four white grapes: assyrtiko, athiri, malagousia, and moscofilero, and two red grapes: xinomavro and agiorgitiko.

Until 2009, 80% of the country's wine production was sold in the domestic market (the Hellenic Statistical Authority (ELSTAT) 2013). This was perhaps unsurprising given that in 2011 only 11% of Greek wine production was of PDO standard (Meloni and Swinnen, 2013), suggesting that much wine was not of a quality to command prices, which justified the cost of export in a very competitive market. The main export markets are the US and Europe mainly Germany which look set to continue to expand. In 2012, there was a 37% rise in Greek wine imports in the US, albeit from a low base, resulting in more than 1000 Greek wines from 90 different wineries being available (ELSTAT, 2013).

However, like many other countries the current world economic crisis has had a severe impact on the country's wine producers and only the wineries who export a significant percentage of their wines have been able to maintain a solid economic basis. This is because not only has the Greek wine industry been hit internally by the recession, but the number of tourists visiting Greece (consuming in bars and taverns) also decreased significantly (Tsartas et al, 2014). Although there was a small increase in incoming tourism after the 2004 Olympics raised the international profile of the country, this was not capitalized upon and the general international tourist profile remains firmly in the declining 4S package (sun, sand, sea, and sex) category rather than in the increasing and more valuable niche tourism market (Tsartas et al, 2014).

The Macedonia Wine Region and Geravassiliou Winery (Ktima)

The Macedonia wine region is one of the oldest in Greece and has a variety of different sub-regions. The most famous of these is Naoussa, which was one of the first AOC regions to be registered (Thalassi, 2010), and is on the slopes of Mount Vermion. Other well-regarded sub regions included Goumenissa and Armindeo (Johnson & Robinson, 2013). In the southern part of Macedonia, closer to the ocean, lies the village of Epanomi, where Ktima Geravassiliou is located. Altogether, it is estimated that there are around 80 wineries in Macedonia (Rynning, 2012).

# of Wineries in Greece	600
# of Wineries in Macedonia	80

The term "Ktima" means "winery" in Greek. Ktima Geravassiliou is located in the village of Epanomi, which is approximately a 5-hour drive north of Athens. The closest city is Thessaloniki, which 25 kilometers north of the winery. The Greeks say that Thessaloniki, the second city of Greece, is where the east meets the west; Classical Greece meets Byzantine and Roman civilization.

The history of the Epanomi settlement goes back almost 6,000 years. Various finds show that vinification was practiced in the area 1,500 years ago, and in Byzantine records Epanomi is described as being a well-known vine-growing region. During the Ottoman Occupation, viticulture and wine-production continued at a lower level and under heavy taxation.

Currently Ktima Gerovassiliou is the only commercial winery in the region. Its nearest commercial neighbor is Tsantali, a large national company with wineries about 30 kilometers distant to the north and east. There is some very small-scale wine production by farmers, sold locally or for home consumption.

Visitor Center at Ktima Gerovassiliou

Vangelis Gerovassiliou started Ktima Gerovassiliou in 1981. Vangelis was born in the village of Epanomi, a descendant of an agricultural family who had lived in the region for generations. After completing a winemaking degree at Bordeaux University, he returned to Greece. He took a position working as winemaker at Chateau Carras for several years before deciding to return to Epanomi to resurrect the family vineyards and build the winery.

Today, Ktima Gerovassiliou encompasses 56 hectares of vineyards as well as a winery and visitor's center. Production ranges from 300,000 to

350,000 bottles (25,000 to 29,000 cases) per year (Papadaki, 2015). The climate is Mediterranean with mild winters and temperate summers, cooled by sea breezes. The vineyard is surrounded by sea on three sides at a distance of three kilometers; on its west side it faces the Thermaikos Gulf and Mount Olympus, which towers over the beaches of Pieria. The soil is mainly sandy with a few clayey substrates and calcareous rocks. It is rich in sea fossils, since the surrounding hilly area was formed by sea deposits (Ktima Gerovassiliou, 2014).

Ktima Gerovassiliou is famous for producing wine from the malagousia grape. This is an ancient Greek variety that Vangelis Gerovassiliou, owner of the winery, is said to have rescued from oblivion. At Ktima Gerovassiliou, Vangelis continues to research and experiment with Greek and foreign varieties. New technological advances blend well with tradition throughout vine growing and vinification processes, however climate changes and the need for sustainable use of natural resources has increased the importance of Greek varieties over international ones because national varieties are genetically better adapted to the local environmental conditions.

The aim of Ktima Gerovassiliou is to produce high quality wines from grapes cultivated exclusively from their vineyard, wines that carry all distinct characteristics of the specific microclimate of Epanomi. Currently the following grape varieties are cultivated: Greek varieties, malagousia, assyrtiko, limnio, mavroudi, and mavrotragano (red). International varieties include: chardonnay, sauvignon blanc, viognier, syrah, and merlot. Currently 30% of the Ktima's production is exported to other European countries, the USA, Canada, Japan, Brazil, Australia, and Singapore. Gerovassiliou wines are also frequent winners in international wine competitions.

Wine Tourism Development in Macedonia and at Ktima Gerovassiliou

Modern wine tourism is a fairly recent phenomenon in Greece. As previously mentioned, because of the image of Greece as a 4S holiday destination and a lack of integrated coordination on the part of the Greek Government (Tsartas et al 2014), tourism in Greece remains dominated by the mass market single destination package model. However, under the EU Rural Development Policy (Corigliano and Mottironi, 2013) the Greek winey industry has benefited from subsidies, which supported the development of wine routes in Greece, including Macedonia.

For Ktima Gerovassiliou the development of the Wine Routes of Northern Greece has been especially beneficial. Ktima Gerovassiliou was one of the founding members of the Wine Producers Association of Macedonia, set up in 1993 as a non-profit making corporation to promote and improve the image of the region's wines. In 2003, they expanded to become the Association of Wine Producers of the Vineyards of Northern Greece and 8 wine routes were developed (Karafolas 2007). Currently 32 wineries are Association members.

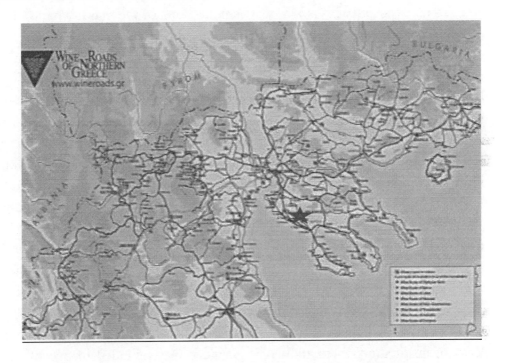

Map of the Wine Roads of Northern Greece and Macedonia

Between the mid 1990s and early 2000s, the Association received substantial funding from both the Organization of Cultural Capital of Thessaloniki (OCCT 1997) and the LEADER II program (Hall et al, 2000: Karafolas 2006). Along with supporting the development of the wine routes these subsidies enabled Vangelis to make the winery more accessible to visitors. For example, they were able to develop their multipurpose visitor center, which includes a wine bar, restaurant, shop, and education area. They also added a museum that was funded privately by the Gerovassiliou family.

THE PROBLEM: WINERY NOT LOCATED ON A MAJOR WINE ROUTE

Despite all of these efforts and the support provided, Vangelis realized that he still had a problem. The location of his winery was isolated, and not located near the other Macedonia wineries. As many wine tourism experts point out, simply making great wine in beautiful natural surroundings is not enough to provide a full wine tourism experience (Hall et al, 2000: Mitchell et al, 2012) especially if your winery is not on one of the major wine routes or wine centers.

When Vangelis started to revive his family vineyard in Epanomi in 1981, he knew that he could make great wines. By opening the Gerovassiliou winery to the public in 1986, he believed that he would also be able to draw tourists to this underdeveloped rural area. Vangelis was aware that sales at the winery and developing future loyal wine consumers were a way of enhancing his business's bottom line. He knew that only relying on a highly competitive international market would always leave his business open to the vagaries of a marketplace he could not control. Therefore, attracting tourists to his winery was very important to his financial success.

However, despite being close to the Greek Riviera and to such cultural icons as the UNESCO world heritage Mount Olympus site, the Macedonia region did not attract as many tourists as other parts of Greece. Furthermore, where he was located, south of Thessaloniki, was not known as a wine tourism destination. Even though his winery was included in one of the 8 wine routes, his route mainly included historical and cultural stops, such as the temple-church of St. George and the wetlands of Fanari with its unique flora and fauna. Therefore, though Vangelis was pleased to be part of the winery association and to receive the benefit of EU funding, he was not receiving the large number of wine visitors that the more famous routes such as Náoussa were. Instead his visitors were primarily Greek day-trippers from Thessaloniki. Vangelis needed to find a way to attract more visitors to his winery.

THE SOLUTION: AWAKEN GREEK INTEREST IN THEIR ANCIENT WINE HERITAGE

Vangelis decided to do some more research on the Greek wine market and consumer behavior. He found that in the past decade more Greeks had started to become interested in quality Greek wine, despite the fact that only 35% of the wine consumed nationally was bottled. He also discovered that Greeks knew very little about their cultural heritage as a great wine producing country.

These findings helped Vangelis realize a potential solution to his wine tourism problem. He decided to attempt to reawaken Greek interest in high quality Greek wines, not only to allow them to rediscover a lost cultural heritage, but also to potentially develop a new future customer base for his own wines. With this in mind, he set about developing a variety of educational events and tools to attract more visitors to his winery, with an emphasis on resurrecting the cultural and historical importance of Greek wine.

Special Events on Greek Wine Culture

Today the winery offers a variety of special events emphasizing Greek wine culture. They organize the "kras tests" (wine tasting events, deriving from the Greek word "krasi", which means wine) for wine lovers and wine experts, plus seminars in order to introduce people to the world and culture of wine. During these sessions they educate visitors on the history and culture of Greek wines, and introduce them to the ancient (almost lost) grape varieties of Greece. They stress the importance of preserving the wine heritage of Greece by resurrecting these ancient varieties and celebrating how unique they are in the global wine world.

The winery also participates in the national Open Cellars events when, for two weekends per year, wineries in Greece are open to the public. During the Day of Wine Tourism each November, visits are free. In addition to the tour of the winery, visitors can watch an historical play related to some aspect of wine culture in the specially constructed theatre at the winery. Concerts are also organized at the winery including a Day of Music, which is also free of charge.

Lectures on Greek Culture

In addition to the above wine related activities, the Ktima Gerovassiliou stages five lectures on culture and art per year to which they invite well known artists, composers, writers, stage directors, poets, actors, etc. During all lectures, guests have the opportunity to taste the Gerovassiliou wines.

Partnership with Universities to Publish Wine Books

Working with Aristotle University, amongst others, each year since 2012, the winery has published a small book on different wine themes. The books always focus on the region of Epanomi, but emphasis different cultural aspects in each volume, such as regional foods. They have also sponsored a children's book that will be made into a film.

As well as forming part of the winery's tourist offering, the aim of the books is to reconfirm the national and international cultural and historical importance of Greek wine to the Greek people. The books are publicized via special events organized at the winery, using social media and the winery's electronic journal.

Interior of the Ktima Gerovassiliou Wine Museum

Greek Wine Museum

Another example of educating visitors about Greek wine heritage is the special museum that has been established at the winery. In 1976, Vangelis started collecting viticultural, winemaking, bottling, and cooperage tools from around the world, including a collection of over 2,600 corkscrews, one of the largest in the world. The collection includes rare and unique pieces dating back to the 18th century, true symbols of the technological advances, high aesthetics, and social structures of the era. The museum is so well regarded; it is featured on the "Taste of Art" website and is included in the list of children's museums of Thessaloniki.

Children's Education Programs on Greek Culture

The overall objective of the educational and cultural activities and events is to promote the culture of wine and Epanomi as a wine and cultural destination, but Vangelis Gerovassiliou and his wife are also significant supporters of local schools and education within the community. As a result, they organize numerous educational programs designed to educate young people on the culture of wine in Greece. The programs they offer have become very popular and are often fully booked out.

Children's Education Program in the Vineyards of Ktima Gerovassiliou

The educational programs are customized based on the age of the audience. For younger children, they present shows and videos, related to wine and gastronomy, but there is no wine tasting. Instead the objective is to introduce them to the winery's processes and place in cultural life past and present. For high school students programs are often based the students' professional career preferences (oenologists, agriculturalists, marketing, export specialists etc.). University students are introduced to aspects such the production, taste and culture of wine, depending on their study interests (agriculturalists, museologist, economists, marketing, etc).

Vangelis, an agriculturalist and oenologist, takes great interest in the visits and often participates in the training sessions. By introducing children to the subject of wine at a young age, the intention is that they will develop a pride in their wine heritage as well perhaps as a future taste for quality wines. This is obviously a long-term promotion strategy as far as Gerovassiliiou wine is concerned, but does address Vangelis passion for reclaiming the lost Greek oenological heritage.

Partnerships with Local Restaurants

Gerovassiliou Winery has also developed partnership with local restaurants. One example is "bring your own wine" day, where customers are encouraged to bring their own wine to local restaurants. The Gerovassiliou team offers wine service training to local restaurateurs and many take them up on the offer. This training has been well received by the younger generation of restaurant owners, who welcome new ideas and suggestions to serving and selling wine. Vangelis is encouraged by this, and has noted that the quality of wine served and the service itself has been slowly but steadily rising throughout Greece.

RESULTS AND BEST PRACTICE IMPLICATIONS

As a result of all these activities the winery has seen their visitor numbers rise, while the average age of their visitors falls, so that they now have to employ two staff members specifically to organize and implement the cultural and educational activities. Currently, the winery receives approximately 10,000 visitors every year (3,000 of which are students). Five years ago the figure was 6,000. The majority of their visitors are Greek, however foreign visitor numbers are increasing. Most come from Ktima

244

Gerovassiliou's major international markets, the US, Canada, and Western Europe (Germany, UK). In the last few years, there has been an increase in "special interest" groups such as dedicated wine tourism groups.

Ktima Gervassiliou Vineyards with Sea in the Distance

Vangelis's strategy to focus on reclaiming and preserving the lost heritage of Greek wines and its unique grape varietals has also paid off in other ways. He has been successful in tapping into the growing international interest in local and indigenous products, by reviving ancient Greek grape varieties. According to Vangelis today, "when we are discussing high quality Greek wine, we are discussing Greek wine varieties". Because of this, his winery has been the subject of many wine journals and tourism articles. This has helped to propel an increase in international tourists.

By applying his expertise in winemaking to craft wines from the heritage Greek grapes, Vangelis has managed to capture many international wine awards. Most recently, his wine was listed in the Top 100 best wines of the world in *Wine Spectator* magazine (2013). Because of this focus, Gerovassiliou wines today are among the most respected Greek wines in the world and export to more than 30 countries. Now his exports have risen from 20% in the past to over 30% today.

The success of Vangelis's strategy has also allowed him to create more high quality jobs and employment opportunities in a rural region of Greece. This supports the Greek Government and EU political goals of increasing employment in Greece, and has brought Ktima Gerovassiliou much praise.

Finally, the winery has been very successful in promoting pride in and consumption of quality Greek wine amongst Greek people. By focusing on reclaiming the lost heritage of high quality Greek wine, Vangelis not only achieved positive acclaim nationally, but also opened the eyes of wine lovers around the world to the glory of ancient grapes and the importance of preserving them for the future.

FUTURE ISSUES

While the Ktima Gerovassiliou is a success story, it cannot stand still if it intends to remain a world-class wine producer and successful independent business. Vangelis needs to continue to expand the image and brand story of Ktima Gerovassiliou to more international markets. He has been successful in Europe, but there is more opportunity to spread the story to the US, Canada, South America, and Asia. Also, because the wine is so unique, it is important to reach out to more highly-educated affluent consumers that are more likely to be interested in buying the wine, rather than mass-market consumers.

Another issue is the challenging economic environment within Greece, and the need for the Greek government to implement a coherent tourism strategy. Though Vangelis has done a good job remaining in touch with political initiatives and potential funding sources, he knows that he will need to remain very involved in the development of tourism in Macedonia, and to continue to encourage the government to stay committed to rural development.

A final challenge is the slow inexorable rate of climate change. In many ways Greece is at the climatic margins of quality wine production. However, the research and experimentation currently being undertaken at the winery, particularly with indigenous varieties, should enable Ktima Gerovassiliou to adapt to many changes, and so remain at the cutting edge of quality wine production for the foreseeable future.

DISCUSSION QUESTIONS

1. How important is the development of the infrastructure in developing agro/wine tourism?
2. How important is passion and commitment in creating a viable business model?

3. Is it ethical to educate school children about a cultural heritage in wine consumption?
4. Does it matter if the wine tourists who visit your winery are national or international?
5. Does the EU Rural Development Policy really achieve its aim of assisting recipients of its grants to maximize their economic potential?

Chapter Fifteen

Luring Wine Tourists to Alentejo Portugal: The Case of Herdade da Malhadinha Nova

Paulo Mendes, *Bordeaux Management School & Vinha Alta*

The Soares family had always dreamed of owning a wine property. As managers of a successful wine distribution business in southern Portugal in the Algarve region, Rita, João and Paulo Soares were very familiar with the diverse wines of their country. So one day in the late 1990's, while traveling further north in the Alentejo wine region of Portugal, they saw a sign that said "For Sale, 300 Acres."

Looking at the landscape, the Soares family thought it seemed perfect for their dream winery. It had beautiful rolling hills and an orchard of olive trees. They made an appointment with the owners of the estate to see the rest of the property and discovered it had enough water and several traditional old houses, though most were very dilapidated. They also learned that the land was ideal to grow grapevines and to raise cattle. After some negotiation they purchased the property that was named "Herdade da Malhadinha Nova", which means "Estate of New Malhadinha."

Paulo and João, along with their spouses, children, and elder parents, then spent several years preparing the land, planting a 50-acre vineyard, retrofitting the old buildings, and designing a new modern winery with tasting room. By 2004, the first wine was released and they began promoting the winery to tourists. However, this was challenging because they were located outside the regulated Alentejo DOC area, and couldn't advertise their wines as higher quality in Portugal. They were also concerned with promoting wine tourism in a sustainable way, and maintaining a balanced lifestyle for their family. João wondered, "We are so involved in this project, including our children; how can we develop it without having 200 strangers invade our house? How do we do this and keep treating visitors as friends?"

OVERVIEW OF PORTUGAL'S WINE INDUSTRY

Portugal is a traditional Old World wine producing country. Its most famous wine region is the Douro, where fortified Port wines are made, and the home of the world's first demarcated DOC (Johnson & Robinson, 2013). Experts believe the grape vine was brought to Southern Portugal by the Phoenicians more than 4000 years ago, and then transported further north by the Romans (Wines of Portugal, 2015).

Today, vines are grown and wines are produced throughout most of the Portuguese territory. In addition to the Douro, other well-known wine regions in Portugal are: Madeira, Dão, Bairrada, and, more recently, Vinho Verde, and Alentejo. Additional production regions include: Lisboa, Tejo, Setúbal, Trás-os-Montes, Beira Interior, Algarve, and Açores. Within each region there are special DOC appellations based on the traditional recognition of quality linked with "terroir".

# of Wineries in Portugal	4000
# of Wineries in Alentejo	300

According to ViniPortugal (2015), the Portuguese Wine Trade Association, wine is a key industry for the Portuguese economy. It represents more than 1.5% of all Portuguese exports, or close to one billion US dollars. Portugal is the 12th largest wine producer and the 9th largest wine exporter in the world. Forty five percent of all wine produced is exported. Historically, Portugal has been one of the countries with the highest wine consumption per capita. In 2014, the country had more than 4,000 registered wineries that produced a total of 6.24 million hectoliter of wine (Instituto do Vinho e da Vinha, 2015).

Portugal is home to more than 120 unique native wine grape varietals (Robinson, 1986). According to ViniPortugal (2015), there are ten flagship Portuguese grape varieties used for the promotion of Portuguese wines. Of these, six are red: 1) Touriga Nacional, 2) Tinta Roriz/Aragonês, 3) Touriga Franca, 4) Trincadeira/Tinta Amarela, 5) Castelão, and 6) Baga, and four are white: 1) Alvarinho, 2) Arinto, 3) Fernão Pires, and 4) Encruzado. Though Portugal is famous for the fortified wines of the Douro and Madeira, today only 12% of production is fortified with the remaining 78% produced as still or sparkling wines (Instituto do Vinho e da Vinha, 2015).

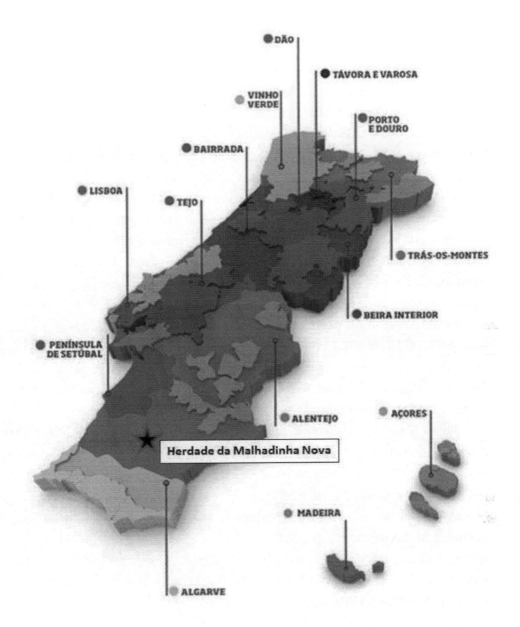

Major Wine Regions of Portugal

Wine Tourism in Portugal

As an important global tourism destination, Portugal is attractive due to its mild climate, nice sandy beaches, safety, and the hospitality of its people. However, when compared with other wine producing countries, Portugal still lags behind in terms of wine tourism development. Until recently, the wine tourism activities were very concentrated in the city of Oporto with most Port producers organizing tastings and visits to the historic lodges in Vila Nova de Gaia. Now with the recent surge in exports and more vigorous international promotion, adventurous wine lovers who are attracted by Portugal diverse wine grapes and styles have targeted the country.

With the gradual increase in wine tourists, the Portuguese wine industry has responded by offering more wine tourism experiences. For example, Douro River cruises to visit family owned wine properties along the river – called quintas - have increased, reaching 600,000 passengers in 2014 (Económico, 2015). In addition, regional authorities and local winery associations have worked hard to implement wine routes throughout the country, although the results vary due to inconsistencies in operation amongst the different wineries. The large distance between wineries, especially in regions like Alentejo, does not help the success of the wine routes.

In recent years, a number of high quality projects were developed, but wine tourism in most parts of the country is still inconsistent. Currently, it is estimated that wineries with structured wine tourism programs range from 40 to 60 (Wines of Portugal, 2015). Given there are more than 4000 wineries in this country; this number is still quite low.

The overall tourism industry in Portugal is very important for the economy and, in 2014, the 16 million visitors generated revenues in excess of 10 billion Euro. Revenues have experienced a CGAR of 8.5% over the last five years (Tourism of Portugal, 2015). Although Tourism of Portugal has identified wine tourism as a strategic segment, no quantitative date is available at the moment.

OVERVIEW OF THE ALENTEJO WINE REGION AND HERDADE DA MALHADINHA NOVA

The Alentejo wine region, where "Herdade da Malhadinha Nova" is located, is in the southeastern part of Portugal bordering Spain. It spans nearly one-third of the country, and is known traditionally for producing cereal and cork. Although Alentejo has been producing wine for centuries, it has only recently emerged as an important wine producing region and wine tourism destination (Tourism of Portugal, 2015).

In 1995, Alentejo ranked as the third largest wine-producing region in Portugal with only 24 independent producers registered and several cooperatives that made wine, mainly for domestic consumption. Today the region is ranked as the largest in wine production with more than 300 wineries, producing close to 1.2 million hectoliters in 2014 (Instituto do Vinho e da Vinha, 2015).

Workers in the Vineyards of Herdade da Malhadinha Nova

Alentejo has differentiated itself by planting some classic grape varietals such as syrah, cabernet sauvignon, and merlot, along with traditional Portuguese varieties. Its warmer Mediterranean climate, which has been compared with the Languedoc region of France, is ideal for producing big reds. Today Alentejo has eight DOC's (now called DOP's), indicating wines with strict rules of control to guarantee authenticity and quality (Wines of Portugal, 2015).

Herdade da Malhadinha Nova is located in the southern party of the Alentejo appellation, close to the town of Beja. From the Lisbon International Airport, it is a two-hour drive to reach the winery resort, and a little over one hour driving time from the Soares family home in Albufeira, a coastal town in the Algarve.

In the 1970's, Paulo and João's parents had started a small wine distribution business called "Garrafeira Soares" in the Algarve near the town of Albufeira. The business grew to become the biggest wine distribution and retail group in the South of Portugal. Therefore by the late 1990's, the family had the resources to buy land, plant a vineyard, and build a winery.

Today Herdade da Malhadinha Nova has a total of 1,130 acres of land with 88 acres of vines and 160 acres of olive trees. Cattle and pigs of "Alentejo Denominated Origin" and "Lusitano" horses are also raised on the property. In addition to the vineyard, farmland and winery, the property also comprises a luxury hotel with ten rooms and spa, a restaurant with a Michelin stared Chef and a horse riding school. The winery produces around 300,000 bottles of wine, mostly red blends (55%), but also white (35%), and rosé (10%).

However, in the beginning, the situation was not as positive as it is today. That's because many international wine buyers and tourists didn't know anything about the wines of Alentejo, and if they did, their perception was usually one of low quality wine. Alentejo was primarily known for producing bread from its vast cereal fields or cork from its many cork trees. Wine from the region generally ended up in inexpensive jug or bulk wine containers. In fact one famous writer referred to Alentejo as the "land of bread and bad wine" (Robinson, 2006).

THE PROBLEM: GARNERING ATTENTION FOR THE UNKNOWN WINES OF ALENTEJO

The Soares family was already very knowledgeable about the wine business and the challenges they would be facing by planting a vineyard in Alentejo and starting their own wine production. Decades in the wine retail business had taught them how difficult it would be to establish a new wine brand.

To compound the problem, Herdade da Malhadinha Nova had the most fantastic soil characteristics to grow grapes, as evidenced by several analysis commissioned by the Soares, but since the property was not within a prized DOC region, their wines would be classified as "Vinho Regional Alentejano," equivalent to "table wine" or "Vin the Pays". They knew their existing distribution and retail business would help push the wines, but it was not sufficient to establish a quality recognized brand.

Rita, Paulo, and João knew they had some challenges to overcome, so they took the time to document them in order to better determine a positive solution:

- *Limited Financial Resources*: A large percentage of the available financial resources were consumed by the purchase of the land and its preparation. Being conservative, the family wanted to limit financial leveraging and make sure the project always had adequate working capital. As João put it "Projects need to be adequately funded and we always keep some cash on hand for difficult times." Therefore, there was not much extra money to spend.
- *Unknown Wine Region*: Alentejo was not yet an important wine producing region and much less a recognized premium wine producing region. Not being placed in a prized DOC area increased the difficulties of pricing the wines at the level they wanted.
- *Marginal Tourism*: In the early 2000's, tourism in Alentejo was a marginal activity very concentrated in its historic capital (Évora) and some beach villages. Wine tourism was almost non-existent.
- *Location Challenges*: Contrary to other wine regions in the World, or even in other parts of Portugal, wineries in Alentejo were too far away from each other and roads were mostly narrow and twisty. To make matters worse, Herdade da Malhadinha Nova was located far from the main wine routes of Évora, Alentejo's capital.
- *Poor Wine Reputation*: Most of Alentejo's wineries, at the time, were volume oriented and mostly cooperatives with focus on production. Few wineries even had a cellar door sales operation.

Instead of demotivating the family, all these challenges acted as an incentive for the whole family to immerse itself in the quest to make their dream happen. Their vision was to become a revered wine producer and acclaimed wine tourism destination.

THE SOLUTION: FOCUS ON QUALITY AND CREATING A WORLD-CLASS WINE TOURISM DESTINATION

Faced with all of the challenges, João recalls his thought at the time. "Portugal is a small country, but the world is so big. If we focus on offering quality and differentiation, we will succeed."

So he decided to travel to most of the major wine production regions in the Old and New World in search of new ideas. He was particularly impressed by the success of wine tourism in California and Australia, and was puzzled by the incipient development of wine tourism in some much older producing regions, particularly in Europe.

From this learning journey, the Soares family drew up a strategy for their winery that would focus on high-end wines produced with the best of both worlds, leveraging the traditional practices of the Old World complemented with the innovations of the New World. They envisioned success by producing the best wines, and then promoting them to select and closely targeted clients. The wines would stand out not only by their intrinsic quality, but also by the consistent, integrated and authentic story behind them - even without the support of a DOC umbrella branding.

Once they had created the highest quality of wine for their terroir, the Soares wanted to establish their winery as a world-class tourism destination. They envisioned sharing their love for wines from their property with guests from around the world. Overtime, they wanted to build a small hotel, restaurant, establish a cattle, horse, and pig farm, and create a variety of engaging wine experiences.

Old Buildings on Property   *New Winery*

Principles for Strategy Implementation & Financing

Once the vision was agreed to, the Soares established a set a guiding principles for strategy implementation. They decided to:

1) Invest wisely; making sure the project generated adequate cash flows for reinvestment. Each portion of the resort would be organized as a separate business unit.

2) Hire a consulting winemaker to produce a range of high quality wines.
3) Leverage wine tourism as a key factor in enhancing the product beyond its intrinsic quality
4) Work with other wineries and local wine promotion organizations to provide a comprehensive and diversified offer.
5) Leverage their existing wine distribution network and retail shops to sell the wine (legal in Portugal), as well as a means to identify select customers.

The project was designed so as each business unit would be self-sustainable. The Soares insisted that no direct cross subsidization was to be allowed between the different business units: farming, winery, hotel, or restaurant. João reported, "Each business area needs to be self-supporting and profitable. This is a philosophy that promotes accountability and responsibility from all business unit managers".

Crafting High Quality Alentejo Wines

The first two harvests were small due to the young age of the vines, and the strict selection of grapes at the vineyard and at the winery. Even though the winery was established as a family operation, the Soares were wise enough to know that they needed to bring in a top-notch consultant to help them produce the highest quality wines.

"This is a family project," explained João Soares. "The business and the family are interconnected. We want to leave our stamp on what we do. Nothing is to be left to chance. But being a family project and trying to be true to ourselves does not mean that it is not professional. We hire only the best for each function. Our winemaking consultant has been the same from day one."

Therefore, they hired Luís Duarte, one of the most renowned consulting winemakers in Portugal, and named winemaker of the year in 2001. He assisted them in designing a winery with the latest equipment, lab, barrel aging facility, and state of the art features. He works closely with Nuno Gonzales, the full-time winemaker at Malhadinha.

Based on their years of experience working in wine retail, the Soares knew they wanted their wines to be in the ultra-premium bracket, and that they wanted to offer three distinct levels. At the time, other Alentejo "quality" producers priced their entry-level wines at $5 to $7 per bottle. Therefore, they positioned their wines as follows:

1) Entry level wine sold under the "Monte da Peceguina" brand retail for between $12 and $25
2) The "Malhadinha" branded wines retail for between $25 and $35+
3) The icon wines are named after the family children – "Marias da Maladinha", "Menino António" and "Pequeno João" - and retail from $35 to $60+.

In the first years of selling the wine, the Soares focused on the on-trade segment and leveraged their distribution and retail business. A number of corporate partnerships were also used to position the wines in the desired ultra-premium segment. They also had a small cellar door operation at the property.

By 2006, the vineyard was already close to full production. It was now time for the Soares to commit some more financial resources and expand their distribution. It was time to start developing the "full fledged" wine tourism project they envisioned since the beginning. This would help position their wines at the top, not only of their region, but also at country and international levels.

Creating a World-Class Wine Tourism Destination

They began by outlining their goals for the wine tourism project:
- To offer customers unique and tailor made experiences
- To price the tourism services in line with the ultra-premium position of the wines
- To only target customers they believe shared their vision
- To leverage their existing winery and ranching operations
- To partner with other local wineries with similar strategies and support the regional development of a broader wine tourism offer

The wine tourism project was developed in the same step-by-step fashion that they had started the winery and ranching operations. Since the acquisition of the 300 acres in 1998, they had remodeled one of the ruined houses and transformed it in the family's house. This was followed by planting the vines, building the winery and tasting room, as well as starting the ranching operations for Alentejo DOP cattle, the famous Portuguese black pigs, and creating a stud farm for Lusitano horses. The first wines were released in 2004.

The next phase for wine tourism was the construction of a luxury design hotel. This would be followed by the addition of a gourmet restaurant, with a Michelin Starred Chef, the following year. In 2006, the hotel plans were laid down and construction followed in 2007. The hotel was ready to be open before Christmas 2007, followed by the restaurant in 2008.

The luxury hotel with ten spacious rooms and suites and spa was erected some distance from the winery to provide quiet and a sense of peace. The architecture follows the traditional local one story construction model. The restaurant was located next to the winery and wine shop so that cellar door visitors could easily get a meal, and hotel guests could leisurely walk over to dine. The restaurant specializes in the local cuisine of the region, and features meals made with the prize Alentejo DOP black pork and local beef, raised on the property.

Time Line of the Project

- 1998 Property with 300 acres acquired
- 1998 200 additional acres acquired
- 1999 Dams for rain water retention built
- 2000 Cattle, horse, and pig production starts
- 2001 50 acres of vines planted
- 2003 Winery built, capacity to process 250 tons
- 2003 First harvest
- 2004 First wine launched
- 2004 Cellar door sales start (wine shop)
- 2006 18 additional acres of vines planted
- 2006 Hotel plans developed
- 2007 Hotel opens
- 2008 Restaurant opens
- 2008 Wine tourism experiences launched
- 2008 630 additional acres of land acquired
- 2010 Winery expanded to handle 400 tons of grapes
- 2012 20 additional acres of vines planted
- 2016 20 new acres of vines and hotel expansion

Wines are sold at the property either to be consumed at the restaurant or taken away. The price of the wines is the same, except for a very small corkage fee if consumed in the restaurant. This does not apply to special releases or library wines that are no longer available in the market, in which case market price is charged.

Creative and Educational Wine Tourism Experiences

In addition, the Soares expanded the wine tourism offerings from traditional tastings and tours at the cellar door and winery to a larger selection of unique wine experiences. Most of these include a stay at the hotel along with meals at the restaurant. Following is a list of these experiences:

- Wine Break – 3 nights relaxing and tasting wine
- Natural Break – 3 days senses test with food and wine
- Mastering Wine – 5 days tasting with expert winemakers
- Mastering Wellness – 5days relaxation with nature and wine tasting
- Sports Master – 5 days with exercise and wine tasting
- Master Dressage – 5 days riding horses and wine tasting
- Photography Break- 3 days taking photos and wine tasting
- Family Break – 3 days with family/children cooking and wine tasting
- Chef Challenge – 5 days with master chefs cooking and wine tasting

In addition to the scheduled wine experiences, Herdade da Malhadinha Nova also offers custom-designed visits. These are prepared by special request and include a large array of options, either from what the property has to offer in wine tastings, horse riding, ATV and all terrain activities, fishing, pigeon shooting, etc., or others outsourced, such as hot air balloon rides. In addition, Malhadinha promotes special events that attract residents from Lisbon and tourists from the most exclusive Algarve resorts like "Quinta do Lago" or "Vale de Lobo". As João, repeatedly, states, "By special request – groups or corporate events – all can be provided, there are no impossibles".

Visitors Enjoying Food and Wine at Malhadinha Resort

Partnerships with Local Wineries and Associations

Another important aspect of their wine tourism strategy was to partner with other local wineries and associations. According to João, "When we started, Alentejo wineries are too far apart from each other, and if you didn't book in advance you may find it closed when you arrive. This is quite different than Napa Valley where you can easily visit 20 wineries within a 20 miles stretch of road."

The Soares were concerned that the lack of alternatives, within a drivable distance of Malhadinha could deter some visitors from driving for two or more hours, not knowing if they would be finding an open tasting room. This was a special concern over the weekends and holidays, when people are more willingly to visit or stay overnight in a wine tourism facility.

So João and his family worked with the closest wineries and with the regional promotion agency, trying to push the development of wine tourism activities in Alentejo. The regional wine promotion agency, Comissão Vitivinícola Regional Alentejana, had launched the "Rota dos Vinhos do Alentejo" (Alentejo Wine Route) in the late 1990's, but its development was gradual. It wasn't until the boom in the number of new wineries of the mid 2000's occurred that the route started attracting more visitors.

As time passed, the Soares were able to successfully align their project with their neighbors and this coincided with the growth in the number of wineries in Alentejo's wine route. Despite some inconsistencies in the services offered and misalignments in individual strategies, visitors now have a much higher probability of finding several wineries open for visits and tastings if they decide to visit without booking.

Lodging and Pool at Malhadinha Resort

Flawless Family-Focused Implementation

Having a clear philosophy and strategy was not enough. Implementing it correctly and consistently was a key issue. The Soares were always well aware of the importance of implementation. As a services business, the "product" is produced and consumed simultaneously, so any failure in the delivery of the service is experienced by the consumer on the spot. No quality control system can allow reprocessing or rejection before the consumer detects it. So, in wine tourism the quality of the people performing the activities is the most important success factor.

By identifying this issue very early, the family members directly involved in the business, recognized they needed to dedicate a disproportionate amount of time to the wine tourism side of the business. Additionally, they recruited and nurtured only the best in their fields. When compared with other tourism business the average education and training level is higher. It was difficult to find people with the right profile: employees that embrace the project's philosophy and could operate with the right discernment and with autonomy. To overcome this debility, the family remained very close to the daily operations, and decisions could be made very quickly, "no corporate ladder or committees to decide" as João explains.

As winemaker Nuno Gonzales explains, "We only hire the best for each function and they also need to share our vision about the project." He quickly adds, "By special request all can be provided to our guest and visitors. Everything is possible. Our staff needs to be aligned with this philosophy of operations".

There are scheduled winery visits and tastings every day, except Sundays. Sunday is the day that the family tries to spend together to relax and reconnect. In this way, they try to preserve some of the family work/life balance that João was concerned about losing when they started the wine tourism experience.

RESULTS AND WHY MALHADINHA IS A BEST PRACTICE IN WINE TOURISM

Herdade da Malhadinah Nova very quickly became a reference in Alentejo. Within a year of opening the hotel, it was recognized as the best wine tourism resort in Portugal, and has received the same award twice sense that time. The venue has been praised in numerous articles in both wine and tourism publications from all over the world, including the *NY Times* and *Conde Nast.*

The wine resort is so popular they sometimes have difficulties in coping with demand. They have been able to achieve prices that most five star hotels in Portugal can only dream of. The rooms sell for $300 per day and, each guest generates, on average, an additional $500 per day.

In terms of wine quality, they have been quite successful at positioning their wines at the ultra premium level and selling at prices that only a few recognized producers can aspire to. They consistently wine awards for their wines, and recently were named by Aníbal Coutinho, a renowned Portuguese wine critic, as the best Portuguese still wine producer in 2014. *Wine Spectator* magazine also has rated some of their wines at 90 points and above (Marcus, 2014).

The strategy adopted by the Soares was successful not only in helping position their wines, produced in a non prized DOC area, but also created a very successful tourism operation. This is much more important, as they are located far from the main traffic routes. When they started, few similar operations existed to create a cluster of wine tourism facilities. The results are much more important as they were able to work, in partnership with other wineries and local authorities, to help promote the whole region. It is certain that wine tourism projects such as Herdade da Malhadinha Nova, along with its neighbor wineries of Herdade dos Grous, Casa de Santa Vitoria, Herdade da Mingorra, and the many other wineries of the Alentejo's wine route were key in putting Alentejo on the world wine tourism map.

In global terms, in almost the same timeframe, Alentejo has evolved from a producer of locally consumed wines to the lead producer of still wines in Portugal, with its wines being acclaimed by national and international critics. In terms of wine tourism, in the last ten years, Alentejo evolved from a non-tourism destination to be named, in 2014, the world's best wine tourism destination by the readers of *USA Today*. In the same year, *National Geographic Travel* magazine included Alentejo in its "Best Trips 2014" list.

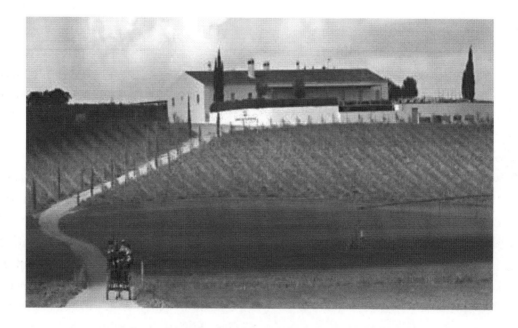

Horse Carriage Ride at Malhadinha Resort

Careful planning, persistent implementation effort, and a clear focus on quality, consistency and distinctiveness are the key success factors of Herdade da Malhadinha Nova. The family resisted the temptation to quit when bad times came or to sway with the winds, and stuck to their plans. As João explained, "The key success factor of Malhadinha has been its careful planning and stage by stage development – in not taking too large a step." He added, "We are entrepreneurs. We are ambitious, but we are also very conservative, and we would never jeopardize the trust our parents deposited in us to develop and manage the business".

FUTURE ISSUES

There are several issues that Malhadinha faces in the future. Up until now, the project was kept at a small size that allowed the family and a team of very aligned and dedicated employees to offer a distinctive service. As the project grows, with another 10 to 20 rooms unit planned to open in the 2016 – 2017 timeframe, the number of professionals will need to grow and the family sense may vanish. As João noted, "The family dedicates a disproportionate amount of time to the tourism area when compared with the revenues generated. All these issues need to be tackled in the coming years as the project evolves, namely by creating a strong culture among their employees, so that the management can become more professional."

Another issue to be taken into consideration is the possibility that the Alentejo wine route may not develop further, and that the wines of Portugal could "fall out of fashion" with obvious repercussions in Alentejo and to Malhadinha. The Soares need to continue working in order to consolidate their positioning and minimize the effects of such an occurrence.

In addition the property is within an environmentally conditioned zone. Today, only 88 acres, of the total of 1,130 acres, are planted with vines. The Soares want to plant an additional 20 acres of new vines increasing the percentage of land occupied by vines to slightly less than 10% but this will certainly involve a challenging negotiation with the zoning authorities. Likewise, global warming experts (Jones, 2006) have identified the Alentejo region as one of the most challenged wine regions from a climate change perspective. If precautions are not taken soon, then this area of the world could be become too hot to grow quality wine grapes.

A final issue, though minor, is that many international tourists have difficulty in pronouncing Portuguese grape vine varietals and winery names. The complete name of Herdade da Malhadinha Nova can be a mouthful for some. Developing a fun and memorable way to pronounce the name correctly for wine tourists could be useful.

DISCUSSION QUESTIONS

1. The Soares followed a strict rule of limiting investments and taking a stage-by-stage approach. Discuss the pros and cons of this strategy and what could be the implications of a more decisive and leveraged approach.

2. The Soares insisted on evaluating each business unit by its own merit and not to consider the impact of cross sales. From graph 1, the wine segment still represents three quarters of the sales volume. Do you think the Soares approach is the most appropriate? How would you account for the contribution of the wine tourism activity in the brand recognition and wine sales?

3. Herdade da Malhadinha Nova is an iconic project, a widely awarded and recognized wine tourism facility at local and international level. If asked to account for its contribution to the development of Alentejo as a wine tourism destination, how would you explain this?

#### Chapter Sixteen

Wine Tourism in a Time of Economic Crisis: The Success Story of Can Bonastre Winery in Spain

Agusti Casas Romeo & Ruben Huertas

University of Barcelona, Spain

It was late September of 2010, after the summer heat had lost its force and the vintage had been harvested, when I first visited Can Bonastre Wine Resort in the Penedes wine region of Spain. It was by chance that I happened upon the winery, because I was following the historical account of Josep Mas i Domènech, a priest and historian who lived in the 1800's, and had written the following lines:

> *"Within the parish of Masquefa, Spain, there are several masia (farmhouses) named Bonastre. Each one takes an adjective to distinguish themselves from each other... the "Hostal del Bonastre", "Bonastre dels Torrents" and "Sta. Magdalena de Bonastre". In the latter, there was a chapel dedicated to St. Magdalena, (which) today no longer exists."*
> (Soler & Valls, 2004).

As a researcher, I was trying to locate properties that appeared in the text, and that was how I encountered the beautiful estate of Can Bonastre Winery, located only 50 kilometers northwest of Barcelona. The term "Can" in Catalan means "someone's home."

Can Bonastre Winery is managed by Gloria Vallés and her brother. When I explained my research project, they kindly invited me in to have a glass of wine. As we talked, they described some of the history of the property, explaining that the first documented reference of the church of Santa Magdalena was from the year 1037 and it was built next to Can

Bonastre's farmhouse. Today the farmhouse is their winery, and when they were renovating it between 2004 and 2007, they found an old will dated 1548. It was written by Joana Mata, who married into the Bonastre family.

As I listened, night fell and we had more wine and some food. Gloria and her brother then began to tell me how business was going. They were worried about the economic hardships and the fallout of the 2008 global recession. Spain had been hit especially hard and tourism was down. The Valles were trying to figure out what actions to take in order to keep the winery and resort going.

Parchment found at Can Bonastre Winery during Renovation

We discussed many options, and I promised to return before Christmas of the next year, but for reasons beyond my control I could not visit again until the end of 2014. This time, the heat of summer was gone, and the cold was intense. We gathered in the house in the early afternoon when the daylight was slowly fading, and had a glass of red wine to warm our palates. Gloria was the only one to greet me this time, as her brother had family obligations with a newly arrived baby. We started talking, and Gloria told me that it had been difficult in the years since my first visit, but they had implemented many of the initiatives we had discussed last time. Now, Can

Bonastre Winery & Resort was not only financially successful, it had become one of the premiere tourist destinations in the region with very unique offerings that included multiple experiential venues for visitors. She went on to explain all the actions they had taken to transform the winery in a time of crisis to become the best practice example it is today.

OVERVIEW OF THE SPANISH WINE INDUSTRY

Spain has been producing wine since Roman times. Today, the Spanish wine industry is a very important and representative industry both nationally and abroad, not only as an income source but also as country image. According to the Ministerio de Medio Ambiente (2010) there are approximately 4,300 wineries in Spain, which represents 14% of the food industry. Furthermore, the Spanish wine industry represents an income of 5,500 million euros, signifying 1% of GDP in 2010.

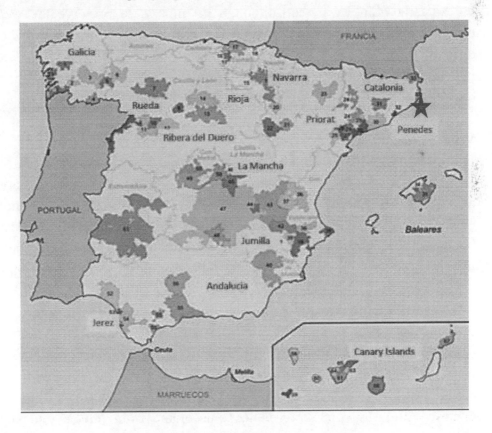

Map of Major Spanish Wine Regions

The Spanish wine industry is organized according to the DO (Domination of Origin). A DO is a governmental recognition that a wine region produces its wines with grapes from the territory and their quality is due to natural factors as soil and weather characteristics from its geographical environment, and human intervention follows the traditional rules. Its main objective is to preserve the high variety and richness of Spanish wines.

There are currently 69 designations of origin (Ministerio de Agricultura, Pesca y Alimentación, 2013). However, among these there are two (La Rioja and Priorat), which have also obtained the distinction of denomination of qualified designation of origin (DOCa.), indicating that the production of quality wines is oldest and most established. In addition, there are two other singularities, DO Cava (sparkling wine similar to champagne) and a new proposition from the Catalan government of DO Catalonia.

This large number of DO has generated an organizational and bureaucratic framework that, although its aims is to preserve the wealth provided by the diversity of wines produced in Spain, it sometimes makes it difficult to market Spanish wine outside of Spain. This is because many international wine consumers are unfamiliar with the intricate system, and only a few knowledgeable consumers understand the DO system.

Wine Tourism in Spain

The starting point for designing wine tourist routes was the designations of origin (DO). The main goal is to integrate, under one thematic concept, environmental resources, and tourist services from a wine region in order to build a differentiated product from other destinations. Other goals include increasing sales of regional products and promoting the socio-economic development of each region.

Each wine route, promoted by the public administration, has a management office to coordinate all business and facilities in its territory. This office also develops strategic plans describing investment needs and their estimated cost, with the aim of to obtaining an economic and sustainable territorial development model over time. Financial resources for operative and strategic performance must come from related companies, public administration, and other institutions.

In the first stage, major investment is needed for creating signage for the wine route and improving infrastructure, such as roads, buildings, and other aesthetics that are important to tourism. In general, public administration funds are used for this. In the next stage, a focus on promotional and marketing materials is necessary, as well as control activities. Here individual wineries and other private companies are involved in the funding.

The Spanish public administration, recognizing the benefits of increased wine tourism, created ACEVIN, the Spanish Association of Wine, to assist in development and design of Spain's Wine Routes. Over the years they, have helped to establish many different wine regions, and today there are 23 wine routes in Spain (ACEVIN, 2015). Despite this positive development, the Spanish wine tourism sector is still in its infancy. This is illustrated by the fact that fewer than 10% of Spanish wineries are open to the public versus up to 80% in Australia and 70% in California's Napa Valley (Marzo-Navarro and Pedraja-Iglesias 2012, 313).

OVERVIEW OF THE PENEDES & CATALONIA WINE REGIONS

Can Bonastre Winery is located in the Penedes wine region of Spain, which is one of the first wine routes to be established in the country. The route is very close to the city of Barcelona, which is the most visited city in Spain near the "golden coast" of the Tarragona beaches. Therefore the region generally attracts many tourists to the wineries, with the latest count at around 435,358 wine tourists (ACEVIN, 2009).

| # of Wineries in Spain | 4300 |
| # of Wineries in Penedes | 221 |

Though Penedes is primarily known for the production of sparkling wine (cava), it also produces some excellent still red and white wines as well. It is the most productive wine region of Spain with more than 26,000 hectares of vineyards and producing 240 million kilos of grapes (INCAVI, 2013). There are more than 14,000 grape growers and 221 wineries (bodegas), including some of the largest in Spain, such as the world-renown Torres family wines and the massive estate of Codorníu.

In order to set up the wine route, it was necessary to develop a communication campaign explaining to the local people the benefits derived from the arrival of tourism. It was also necessary to organize training courses to improve the professional skills of human resources and, also, creating a Tourism Observatory with the objective to study and make forecasting about supply and demand for the product-tourist destination. Although there is a long tradition of wine production in Spain, wine tourism was a novelty (Gazquez-Abad *et al.*, 2015).

Though Can Bonastre is located within the territorial area of the Penedès, it also belongs to the DO Catalonia, which in practical terms takes its place in an "umbrella brand". This wine producing area has a Mediterranean climate, characterized by having many hours of sunshine (> 2500 h / year), with dry and mild winters, and not too hot summers. Spring and fall are usually the most unstable and rainy seasons. The average annual temperature is around 14-15°C and rainfall ranges from 350 mm / year in the drier to over 600 mm / year in the wettest areas.

The predominant soils are sedimentary deposits of calcareous nature, poor in organic matter and medium texture. This DO Catalonia has a planted area of 3,000 vines/ha. The maximum output allowed to produce wines is 10,000kg/ha for red grape varieties and 12,000 kg/ha for white grape varieties.

Panoramic View of Can Bonastre Wine Resort

DO Catalonia Wines cover white, rosé, and red wines. It also produces "crianza", "reserve", and "gran reserve", with a time in barrel, in the case of red wines, 6, 12, or 24 months a total aging period of 2, 3, or 5 years respectively. Traditional vine varieties are often mixed with foreign varieties recently implanted to produce excellent wines. In total, the DO Catalonia has approved 35 varieties of grapes for their wines.

THE PROBLEM: CAN BONASTRE WINERY IS HIT BY THE ECONOMIC CRISIS

When the wine tourist route was first established in Penedes between 2001 and 2006, the number of wineries increased dramatically. This growth was encouraged by ACEVIN, because according to their strategic plan, tourism was expected to increase in this region. It was an optimistic period, the Spanish economy was growing, and all forecasts suggested the growth would be sustainable. Therefore many, wineries, including Can Bonastre, invested in building new spaces to receive tourists and adapting old spaces to new quality standards required by ACEVIN.

View of Can Bonastre Spa with Mountains

The Valles had been operating Can Bonsatre as a winery since 1996 on 100 hectares of land, producing 13 different varieties, including pinot noir, cabernet sauvignon, and merlot. The winery had a long history of receiving numerous national and international quality awards. Then, in April of 2007, with the encouragement of ACEVIN, they added a 12 room hotel, restaurant, and spa, and renamed themselves Can Bonastre Wine Resort.

However, the positive tourism forecasts were not accurate, and after the global financial crash of 2008, Spain went into a deep recession. The government drastically reduced its investment in new infrastructure and tourism promotion. In addition, financial institutions, due to their own difficulties restricted lending money to small and medium firms. By 2010, the Spanish economic crisis had worsened and all product and services lines were affected.

Gloria Valles and her brother Roger at Can Bonstrare Winery were very concerned. There was a widespread perception among Spanish consumers that they faced a prolonged crisis and it would take long time to recover their economical situation. Luxury spending on activities such as tourism decreased, and since Can Bonastre's market was primarily local tourists, the impact was immediate and painful. The fact that their business was also seasonal, due to the fact that many tourists visited the beaches of Barcelona in the summer, and therefore winters were much slower, made the situation even worse.

Financial pressures mounted for the winery, with suppliers and finance agencies requesting payment. Borrowing more money was difficult because banks, mostly indebted, were trying to solve their own problems by reducing credit to their customers. The situation worsened when competitor wineries started cutting prices to stay in business, which created more financial pressure for Can Bonastre.

The situation was bleak. What to do? Where to begin? The only way not to go into receivership was to reduce costs. But how? Gloria and Roger Valles could not waste a minute. Their resort expansion, which they had launched with great enthusiasm, was at stake. They needed to take drastic action to save their wine business.

SOLUTION: CREATIVE OUT OF THE BOX THINKING FOR EXPERIENTIAL WINE TOURISM

They began brainstorming different options and scenarios, some which focused on reducing costs and others, which stimulated demand. In the end, it was their creative "out of the box thinking" about wine tourism in Spain that allowed them to turn around the situation in a couple of years and become successful. This began when they started researching what had worked well in other international wine regions and learned about the concept of "experiential wine tourism."

Experiential wine tourism is providing a unique experience, which includes not only wine tasting, but a connection to gastronomy, art, culture, education, and travel (Hall & Mitchell, 2000; Getz & Brown, 2006; Marzo-Navarro & Pedraja-Iglesias, 2009; Scherrer et al, 2009; Cohen & Ben-Nun, 2009; Dougherty, 2012). Traditionally, in the Penedes wine region of Spain, wine tourism had focused on the standard wine tasting and tour of the winery facilities. In general, the majority of the visitors to Can Bonastre were from Spain, with very few international visitors. Also, the fact that Can Bonastre was located near Barcelona, generally meant that tourists first visited the larger more famous wineries of Torres, Codorniu and Freixenet. Then if they had time, they managed to make their way to Can Bonastre.

Therefore, Gloria and Roger decided to create a point of differentiation and focus on attracting international wine tourists. This turned out to be a good strategy, because the rest of Europe began to recover from the economic crisis earlier than Spain. Visitors from Germany and England who appreciated Spanish wine as well as the emerging middle class in Eastern Europe became their market focus as they restructured their marketing strategy and experiential portfolio of services in the following ways:

Re-focus Portfolio on Weddings and Business Meetings

In their analysis of the current situation and the assets they had which could be useful in generating revenues, they realized that the restaurant and hotel linked with the winery could be quite attractive to the more lucrative market segment of weddings and business meetings. Also the fact that they were only 50 kilometers (31 miles) from Barcelona gave them easy access to the international airport. Their beautiful grounds with vineyards, lake, and mountains in the background were very scenic. Therefore, they re-designed their marketing materials and website to emphasize a focus on weddings and business meetings – a new type of experiential wine tourism.

Reduce Costs by Limiting Wine Tourism Hours

In order to reduce costs, they decided to reduce the number of hours the winery, hotel, and restaurant were open to tourists. Therefore, instead of being open every day, they changed the resort to only being open on the weekends, or for private events, such as the weddings and business meetings. Though this required that they reduce staff, it allowed them to decrease many overhead and energy costs. As the new focus on private weddings and business meetings took off, they were able to bring back staff on an as needed basis.

Wedding at Can Bonastre Wine Resort

Implement a Digital Marketing Strategy

Another action they took was to implement a digital marketing strategy. They updated their website to show all of their offerings in both Spanish and English. The menu begins with "Events and Meetings," and then moves onto "Weddings, Hotel, Restaurant, Spa, Winery." In this way, they emphasize the many different experiential offerings of the property. The website was professionally designed to appeal to international tourists, and also includes a blog and videos of past events. Most importantly, they invested in expanded search engine optimization (SEO), so that their website would show up in more search engines.

Expand Marketing Message to Include DO Catalonia

The Valles recognized that being part of the DO Penedes region, though useful, caused some confusion for visitors who were not familiar with the complicated DO system. Also, the fact that they were not a famous cava producer, and focused more on red wine, caused them to be lost in the shuffle of the giant brands of Torres, Codorniu, and Freixenet. Therefore, they decided to reposition themselves as part of the greater DO Catalonia region, because it was more oriented to foreign markets. Indeed the tourism arm of Catalonia was more aggressive in marketing multiple regional products, including food, culture, and wine. This fit better with the updated marketing strategy of Can Bonastre, and helped more international tourists find them.

Company Teambuilding Event

Implement a Strategy of Dramatic Energy Savings

After some research, the Valles also decided to adopt alternative energy solutions to save money. They invested in biomass generator power equipment to recycle vines pruning remains and other debris generated by garden maintenance. In addition, they installed low-energy lighting through the resort and winery to reduce energy costs. Though these changes required an upfront investment, in the long run it has reduced their energy consumption bill dramatically.

RESULTS – A TURN AROUND SUCCESS IN WINE TOURISM

The story of how Can Bonastre Wine Resort survived the drastic economic recession of Spain and implemented a turn-around strategy can definitely be considered a best practice in wine tourism. Though encouraged by the Spanish government to invest in tourism, no one could have predicted the devastating impact on Spain of the global recession.

When the crisis hit, the number of tourists visiting Spanish wine regions declined overnight, and the seasonal fluctuation was more marked. Spain's economic crisis also took longer to end than the rest of Europe, and therefore focusing on traditional Spanish wine tourists was not the answer. Can Bonastre owners exhibited resiliency and innovation by analyzing the situation and developing a set of solutions to both reduce costs and increase demand, thus creating a new type of experiential wine tourism in the Catalonia/Penedes region.

The results of their efforts speak clearly to a wine tourism best practice. By restructuring their services to focus on weddings and business meetings, and implementing a digital market strategy, they were able to book 40 weddings and more than 50 meetings in the first two years of the implementation. Since that time they have booked even more events, including such famous companies as Mercedes-Benz, Land Rover, Swatch, and Hanos from Switzerland.

In terms of social media, they have very high reviews on TripAdvisor. Of the more than 85 tourist reviews, 76% are excellent or very good. Two recent reviews stated:

"Blissful (5 star review). "I absolutely loved this place… our stay was 100% enjoyable and relaxing. There was a wedding on

one of the days we were there, but due to the location of the function room this had absolutely no effect on us - no disturbance."

"Heaven on Earth (5 star review) "A must do for winelovers and people who appreciate amazing views in combination with wellness facilities. Excellent service, good food, and extremely friendly staff. The wines of Canbonastre are very tasteful, excellent price/quality rapport. This pearl is certainly a place to go back."

A more current example of success at Can Bonastre is that the resort has been selected as a great location for filming of TV commercials. For example, Mazda has completed several commercials for their latest vehicles there, and in 2015, the famous singer Shakira, shot two ORAL-B commercials at Can Bonastre Wine Resort (Oral-B, 2015).

FUTURE ISSUES

Despite all of its successes, Can Bonastre Wine Resort still faces several issues. They are still working at reducing the debt overhang by focusing the business exclusively on the most profitable segments. However, it will still take a number of years before they are in a comfortable financial position. They continue to look for cost cutting actions that will not hurt quality or service, such as hiring talented staff that add value.

They are facing increasing competition from copycat competitors who offer more aggressive pricings and counteroffers adapted to local and international calendars. Therefore, they have to continue to innovate, and focus on offering the best customer service and experience for their guests.

Though they have been successful with their digital marketing strategy, they recognize they need to do more to increase visibility and brand awareness. They need to expand their SEM and social media presence online and with mobile technologies.

They are also considering the idea of doing a "Joint Venture" with another winery to produce more high quality wines. They believe this will allow them to new segments of international customers who are interested in wine at different price points.

DISCUSSION QUESTIONS

1. In Spain, there is a tradition of excessive government intervention in the economy. This situation has created a management culture of follow the government actions. The crisis has taught companies some lessons, one of them is to step back and gain perspective of government indications. Since the recession began, there has been a precipitous decline in trust, and many companies were forced into bankruptcy. What should Spanish wineries and other business do to prevent this from happening again in the future?

2. Conduct a SWOT analysis on Can Bonastre Wine Resort now that it has been restructured.

3. Identify a list of new experiential wine tourism activities that Can Bonastre Wine Resort can offer to attract both local and international tourist?

4. What should Can Bonastre Wine Resort do to enhance its digital marketing strategy to enhance revenues?

5. What should Can Bonastre Wine Resort do to expand its energy conservation strategy and reduce costs?

ABOUT THE EDITORS

Dr. Liz Thach, MW (pronounced "Tosh") is the Distinguished Professor of Wine and a Professor of Management at Sonoma State University where she teaches in both the undergraduate and Wine MBA programs. Liz's passion is wine, and she has visited most of the major wine regions of the world and more than 37 countries. In addition, she is an award winning author who has published over 120 articles and 6 wine books, including *Call of the Vine* and *Wine – A Global Business*. A fifth generation Californian, Liz finished her Ph.D. at Texas A&M and now lives on Sonoma Mountain where she tends a small hobby vineyard and makes pinot noir wine. She also works as a wine judge in various competitions, and has served on many wine editorial and non-profit boards. Liz obtained the distinction of Master of Wine (MW) in May of 2011.

Dr. Steve Charters, MW is Director of Research in the School of Wine and Spirits Business, ESC Dijon/Burgundy School of Business in Dijon, France. He was previously Professor of Champagne Management at Reims Management School and before that taught at Edith Cowan University in Perth, Australia. He develops research projects and courses relevant to the wine business. His research interests include the relationship of wine to place, drinker perceptions of quality in wine, the mythology surrounding wine consumption, the motivation to drink, and wine tourism. He is a member of the editorial board of the *International Journal of Wine Business Research, the Journal of Wine Research* and the *British Food Journal* and is one of only 325 members of the Institute of Masters of Wine in the world.

ASSISTANT EDITORS

Ms. Miranda Hardison is an English and Liberal Studies student at Sonoma State University in the Hutchins School. She is passionate about spelling and grammar, and worked as a part-time editing assistant on this book. Her previous work experience included being editor for her high school yearbook and school newspaper, as well as working as an editing intern at the *Orange County Register*.

Ms. Kendall Cavanaugh is a Payroll & Accounts Payable Specialist with Marr B. Olsen Incorporated in the San Francisco Bay Area. She has a love of fine wine and cuisine, and obtained her B.S. degree at Sonoma State University in Business Administration with a concentration in Wine Business. While completing her university classes, she worked part-time as a wine tourism research intern and assisted with research and editing for this book. She also created several of the wine region maps.

Ms. Chelsea Scherer is a Distributor Rep with Young's Market Company in California doing wine sales and merchandising. She has a passion for wine and obtained her B.S. degree at Sonoma State University in Business Administration with a concentration in Wine Business. While completing her university classes, she worked part-time as a wine tourism research intern and assisted with research and editing for this book.

AUTHOR BIOS BY CHAPTER

Chapter 1 - Introduction

Written by the Editors

Chapter 2 – Austria

Dr. Albert Stöckl is programme director of the Bachelor programme "International Wine Business" at the IMC Krems University of Applied Sciences, Austria. He spent eight years working and studying in Germany, France, Great Britain, Italy, and Sweden and is specialized on wine tourism and related research topics.

Dr. Cornelia Caseau is head of the Department of Languages and Cultures at Burgundy School of Business, Dijon, France. She originates from Vienna and among other areas she does research on German and Austrian culture, literature, society, and economy.

Chapter 3 – Canada

Dr. Linda Bramble is a business consultant, executive leadership coach and wine writer. She obtained her Ph.D. in philosophy and education from the University at Buffalo and subsequently was on the faculties at the universities of Concordia (Montreal) and Brock (St. Catharines, Ontario). She is a certified sommelier and taught courses for the Canadian Association of Professional Sommeliers. She has written several books on wine touring, winery leadership, and the history of the Ontario wine industry.

Dr. Carman Cullen is an Associate Professor of Marketing specializing in wine marketing for the past several years. He is on the faculty of the Goodman School of Business in Brock University, Canada. Carman is the recipient of numerous teaching awards and has offered seminars around the world.

Chapter 4 – China

Ms. Wenxiao Zhang is the Deputy Secretary for the International Federation of Vine and Wine of Helan Mountain's East Foothill, located in Yinchuan, Ningxia, China. She studied wine business at the Bordeaux Management School in France, and currently assists with marketing strategy and public relations for the Ningxia Wine Region.

Dr. Liz Thach, MW is the Distinguished Professor of Wine at Sonoma State University in Rohnert Park, California, USA. She teaches in both the undergraduate and Wine MBA programs at the Wine Business Institute on campus. She obtained her Ph.D. at Texas A&M University and the distinction of Master of Wine (MW) in 2011.

Chapter 5 - Bordeaux

Juliette Passebois Ducros is associate professor at IAE Bordeaux University School of Management and is a member of the marketing research team of IRGO center of research. She holds her PhD in Management obtained at University of Montpellier in 2003. She teaches marketing, consumer behavior, and data analysis. Her research addresses issues related to marketing in museums and consumer experiences with art and culture.

Julien Cusin is Associate Professor at the IAE School of Management (Bordeaux University, erm/IRGO). He teaches Strategic Management, Human Resource Management and Knowledge Management. He also runs a Master in Human Resource Management. His main area of research is management of failures both at organizational and individual levels.

Chapter 5 – Burgundy

Ms. Laurence Cogan-Marie has recently joined the School of Wine and Spirits Business at ESC Dijon where she teaches wine-tourism and culinary tourism. Her expertise relates more particularly to the marketing of wine, and the situation of wine-tourism in Burgundy and in Beaujolais. She is a graduate of ESCP Europe and holds an MSC in in International Business and a Diplomkauffrau.

Dr. Steve Charters, MW is Director of Research in the School of Wine and Spirits Business, ESC Dijon/Burgundy School of Business in Dijon, France. He was previously Professor of Champagne Management at Reims Management School and before that taught in Perth, Australia, where he gained a PhD from Edith Cowan University.

Dr. Jo Fountain is a Senior Lecturer in Tourism Management at Lincoln University, New Zealand. Her research interest in wine tourism dates back more than a decade and encompasses a range of contexts, including New Zealand, Australia and France. She is interested in visitors' experiences of wine tourism, and also the impact of wine tourism on future wine consumption behavior. Particular market segments have been a specific focus of much of this research, with early studies focused on Generation Y, and more recent projects exploring the Chinese market as wine consumers and wine tourists.

Dr. Claude Chapuis is Professor in the School of Wine and Spirits Business, ESC Dijon/Burgundy School of Business in Dijon, France. Expert in viticulture and many topics linked to wine and culture. He worked many years as winegrowers in Burgundy, California, Australia, New Zealand, South Africa, and in Mosel Valley in Germany. He wrote many books and articles on wine with a specific focus on History, viticulture, and Burgundy.

Dr. Benoît Lecat is Head of the Wine and Viticulture Department at California Polytechnic State University in the USA. Previously he was Professor of Marketing in the School of Wine and Spirits Business, ESC Dijon/Burgundy School of Business in Dijon, France. His research has focused on management of luxury goods, fine wines, and spirits with a specific interest on price premium, price determinants, m-commerce, scarcity, terroir, and tourism.

Chapter 7 – Provence

Dr. Coralie Haller is a wine management and information system associate professor at EM Strasbourg Business School. She obtained her PhD at Aix Marseille University in 2014. Her research interests concern management of information in SME in the wine industry. She is head of the Major in Wine Management and Tourism and director of the Executive Education at EM Strasbourg Business School.

Dr. Bédé Sébastien is associate professor in strategy at the EM Strasbourg Business School in France, and researcher at the Humans and Management in Society (Humanis) research center. His primary research interests focus on governance of tourism destination. His research fields concern remembrance tourism and wine tourism. He completed his PhD at Nice Sophia Antipolis University in 2013.

Dr. Michel Couderc, agro-oenologist, economist is in charge of Economy, Studies and Strategy, and develop the World Rosé Observatory at CIVP. He takes part in numbers of information diffusion, scientifically or general public editions or conferences.

Mr. François Millo, is General Director and President of the Provence Wine Council (CIVP). He has multiples activities linked to wine, including the publication of a new book, *Provence Food and Wine: The Art of Living.* He is also an agronomic engineer and a professional photographer.

Chapter 8 – Italy

Dr. Roberta Capitello, Ph.D., is Associate Professor in Agricultural Economics at the Department of Business Administration, University of Verona. She teaches Wine Economics and Marketing. Her research activity is focused on consumer behavior and food and wine marketing.

Dr. Lara Agnoli, Ph.D., is Research Fellow in Agricultural Economics at the Department of Business Administration, University of Verona. Her research activity is focused on demand analysis, wine tourism, and food and wine marketing.

Dr. Ilenia Confente, Ph.D., is Assistant Professor in Marketing and Supply Chain Management at the Department of Business Administration, University of Verona. She teaches Advanced Tourism Marketing. Her research activity is focused on web marketing and business to business marketing.

Dr. Paolo Benvenuti is Director of Associazione Nazionale Città del Vino.

Dr. Iole Piscolla is a journalist and an expert in wine tourism technical and territorial economic development projects.

Chapter 9 – New Zealand

Dr. Lucy Baragwanath manages the Auckland Council Research and Evaluation Unit. She has a varied background having worked in various public and private sector roles as a researcher, advisor, analyst, and strategist. She has a PhD from Lincoln University and undertook postdoctoral study at Lancaster University and the University of Auckland. She has a strong interest in stakeholder engagement, strategic planning, and economic development.

Dr. Nick Lewis is an economic geographer who has studied the development of New Zealand's wine economy for the last fifteen years. He has published academic papers on questions of governance, terroir, and mobilizing representations of place to create economic value. He has also worked with wine industry organizations and national economic development agencies to develop regional development strategies.

Chapter 10 – Sonoma, USA

Dr. Liz Thach, MW is the Distinguished Professor of Wine at Sonoma State University in Rohnert Park, California, USA. She teaches in both the undergraduate and Wine MBA programs at the Wine Business Institute on campus. She obtained her Ph.D. at Texas A&M University and the distinction of Master of Wine (MW) in 2011.

Ms. Michelle Mozell is a Research Associate with Wine Opinions, a wine analytics firm headquartered in California. She also works part-time as a wine writer and researcher. She holds an MBA in Wine Business from Sonoma State University. Her current academic focus is the impact of climate change on the global wine industry and the wine and wine tourism industry of Sonoma County, California.

Chapter 11 – Argentina

Dr. Jimena Estrella Orrego is assistant researcher at Universidad Nacional de Cuyo, Faculty of Agrarian Sciences. After her master degree at Udine University she gained her PhD in Agricultural Economics at Padova University, Italy.

Dr. Alejandro Gennari is full professor at Universidad Nacional de Cuyo, Faculty of Agrarian Sciences. He teaches annually at Vinifera Montpellier and numerous postgraduate courses. He was previously Executive Chief at the Irrigation Organization from Mendoza.

Chapter 12 – Tasmania, Australia

Dr. Marlene A. Pratt is a lecturer within the Department of Tourism, Sport, and Hotel Management at Griffith University on the Gold Coast, Australia. Marlene's research interests include wine tourism, consumer behavior, events, and experiential learning. Marlene completed her PhD in 2011 on wine tourism focusing on understanding visitors' intentions and motivations to visit wine regions.

Chapter 13 – Beaujolais

Dr. Joanna Fountain is a Senior Lecturer in Tourism Management at Lincoln University, New Zealand. Her research interest in wine tourism dates back more than a decade and encompasses a range of contexts, including New Zealand,

Australia, and France. She is interested in visitors' experiences of wine tourism, and also the impact of wine tourism on future wine consumption behavior. Particular market segments have been a specific focus of much of this research, with early studies focused on Generation Y, and more recent projects exploring the Chinese market as wine consumers and wine tourists.

Ms. Laurence Cogan-Marie has recently joined the School of Wine and Spirits Business at ESC Dijon where she teaches wine-tourism and culinary tourism. Her expertise relates more particularly to the marketing of wine, and the situation of wine-tourism in Burgundy and in Beaujolais. She is a graduate of ESCP Europe and holds an MSC in International Business and a Diplom-Kauffrau.

Chapter 14 – Greece

Dr. Konstantinos Rotsios is the Associate Dean of Perrotis College at the American Farm School in Thessaloniki, Greece. Konstantinos teaches marketing related modules. His current research interests include Food SME's and rural development, food exports, and Greek SMEs and more specifically wine and olive oil exports and export success factors. Konstantinos has coordinated and participated in many EU and national projects related to Rural Development and Food SMEs. His work has been presented in numerous conferences in Greece and abroad.

Dr. Caroline Ritchie is a Reader in Hospitality. She is based in the Welsh Center for Tourism Research at Cardiff Metropolitan University, Cardiff, UK. Her PhD identified the profile and attitudes of UK wine consumers. Caroline teaches in the area of hospitality, specifically food and beverages, and on wine based programs such as those of the Wine and Spirit Education Trust. Her research interests include socio/cultural interaction with wine and its impact on consumer behavior She has been involved in several national and international research projects in these areas.

Chapter 15 – Portugal

Mr. Paulo Mendes is the Managing Partner of Vinha Alta, a wine business boutique consultancy. Before starting his own consulting firm, he founded and was the CEO of Madeira Vintners, the only Madeira wine producer launched on the island in over half a century. Previously, he was a consultant for Arthur D. Little in several European and American countries and held several management positions in Portuguese and international companies. He has a Wine MBA from the Bordeaux Management School. He currently works and lives in Lisbon, Portugal.

Chapter 16 – Spain

Dr. Agusti Casas-Romeo is associate professor of marketing at the Faculty of Economics and Business, University of Barcelona (Spain). He investigates marketing issues, especially in market research, and the environment Wine (wine management and sustainability). His works have been published in international magazines such as *Cornell Hospitality Quarterly* and *South African Journal of Business Management.* He has been awarded the Ruth Green Award, prize for the best case brought by authors outside the USA at the NACRA conference.

Dr. Ruben Huerta-Garcia is associate professor of marketing at the Faculty of Economics and Business, University of Barcelona (Spain). His research interests cover several marketing topics, especially those related to marketing research, web design and Internet research, and tourism management. His work has been published in international journals such as *International Journal of Tourism Research* and *Journal of Business & Industrial Marketing* among others. He has also contributed to more than 80 papers to conference proceedings.

References & Photo/Map Credits
by Chapter

Chapter 1: Introduction and Overview of Wine Tourism by Steve Charters & Liz Thach

Photo Credits
Photo of wine tourists, purchased from Veer.com, BLP0011526

References
Asian Wine Association. (2015). Charter. *Asian Wine Association web*site. Retrieved on 9/17/15 at http://asianwinesassociation.com/charter/

Bogan, C.E. and English, M.J., Best Practices, LLC (1994). *Benchmarking for Best Practices: Winning Through Innovative Adaptation*. New York: McGraw-Hill.

Bruwer, J., Pratt, M. A., Saliba, A., & Hirche, M. (2014). Regional destination image perception of tourists within a winescape context. *Current Issues in Tourism*, 1-21.

Business Dictionary (2015). Definition of Best Practices. *BusinessDictionary.com*. Retrieved on 9/18/15 at http://www.businessdictionary.com/definition/best-practice.html

Carlsen, J. (2004). A review of global wine tourism research, *Journal of Wine Research*, 15(1), 5-13.

Carlson, J. & Charters, S. (2006), editors. *Global Wine Tourism*. London: CABI.

Charters, S. (2010). New World and Mediterranean wine tourism: A comparative analysis, *Tourism*, 57(4), 369-379.

Charters & Spielman (2014). The Characteristics of Strong Territorial Brands: the Case of Champagne. *Journal of Business Research*, June 2014.

Correia, L. Passos A., & Charters, S. (2004). Wine routes in Portugal: A case study of the Bairrada Wine Route. *Journal of wine research*, 15(1), 15-25.

Gilmore, J. H., & Pine, B. J. (2007). Authenticity: What consumers really want. *Harvard Business Press.*

Menival, D., & Charters, S. (2011). The impact of tourism on the willingness to pay for a bottle of standard quality champagne (No. 01-2008). *Enometrica-Review of the Vineyard Data Quantification Society* (VDQS) and the European Association of Wine Economists (EuAWE)-Macerata University, Faculty of Communications.

Dodd, T.M. (1995) "Opportunities and Pitfalls of Tourism in a Developing Wine Industry", *International Journal of Wine Marketing*, Vol. 7 Iss: 1, pp.5 - 16

Getz, D. (2000). *Explore wine tourism: management, development & destinations*. NY: Cognizant Communication Corporation.

Hall, C.M. (1996). Wine tourism in New Zealand. In G. Kearsley (Eds.) Proceedings of Tourism Down Under II: A Tourism Research Conference, Dunedin: Centre for Tourism, University of Otago.

Hall, C.M. & Macionis, N. (1998). Wine tourism in Australia and New Zealand. In R. Butler, M. Hall and J. Jeckins (Eds.) Tourism and Recreation in Rural Areas, England: Wiley.

Hall, C. M., Johnson, G., Cambourne, B., Macionis, N., Mitchell, R., & Sharples, L. (2000). Wine Tourism: an introduction, In C. M. Hall, L. Sharples, B. Cambourne, N. Macionis, R. Mitchell & G. Johnson (Eds.), *Wine Tourism Around the World: Development, management and markets* (pp. 1-23), Elsevier Science, Oxford.

Lee,K. (2015), Editor. *Strategic Winery Tourism and Management: Building Competitive Winery Tourism and Winery Management Strategy*. VA: Apple Academic Press

New Zealand Tourism (2014) *Tourism Profile: Special Interest Tourism - Wine Tourism*. From http://www.tourismnewzealand.com/media/1132209/wine-tourism-profile.pdf Accessed 26th March 2015

O'Neill, M., & Charters, S. (2000). Service quality at the cellar door: implications for Western Australia's developing wine tourism industry. *Managing Service Quality: An International Journal*, 10(2), 112-122.

SA Tourism Industry Council (2009) *South Australian Food and Wine Tourism Strategy*, From http://satic.com.au/images/uploads/documents/foodwine%20tourism%20strategy.pdf Accessed 21st January 2014.

Thach, L. (2013, January). Twelve best practices in global wine tourism. *Fine Wine and Liquor Magazine, 71,* 44-47.

Thach, L. (2007). "Trends in wine tourism." *Wine Business Monthly* 15: 2007.

Thach, E. C., & Olsen, J. (2006). The role of service quality in influencing brand attachments at winery visitor centers. *Journal of Quality Assurance in Hospitality & Tourism*, 7(3), 59-77.

Chapter 2: AUSTRIA - Luring Tourists Back to the Traditional Wine Taverns (Heurigens) of Vienna, Austria by Albert Stöckl & Cornelia Caseau

Map and Photo Credits:
Map of Austrian Wine courtesy of the Austrian Wine Marketing Board
Photos of WienWein Logos and Six Men courtesy of WienWein
Photos of Heurigens courtesy Danube Tourism Board and Austrian Wine Marketing Board

References
Austrian Wine. (2015). Wines of Austria website. Retrieved on Aug. 4, 2015 at http://www.austrianwine.com/.

Arbeithuber, B., Waxenegger, B., & Skurnik, K. (2011). Documentation 2011.

Structure Wine Country Austria. Austrian Wine Marketing Board, Wien.

Bachmayer, R. (2012). Der Heurige als Kulturgut im Wandel der Zeit. Diplomarbeit Weinakademie Rust.

Baumgartner, G. (2004). Heurigenkultur und Weinbau in Grinzing. Aktuelle Situation und Analyse der Buschenschanken und ihre Stellung gegenüber Heurigenrestaurants. Projektarbeit am Österreichischen Universitätslehrgang für Tourismuswirtschaft an der WU Wien.

Beiglböck, B. (2007). Kundenstrukturerhebung in südburgenländischen Buschenschenken und marketingpolitische Empfehlungen, Master Thesis, Burgenland University of Applied Sciences.

Gross, M. (2014). Within Vienna's city limits, some serious wines are being made. *NY Post*, Aug. 30, 2014. Retrieved on 8/4/15 at http://nypost.com/2014/08/30/within-viennas-city-limits-some-serious-wines-are-being-made/

Keen, M. F. & Robinson, G. (2001). The Viennese Heuriger: Cultural Resilience and Commercialization amidst the Vineyards, *The Journal of Popular Cul*ture, Vol. 35, Issue 1, 219–234.

Kopsitsch, K. (2008). Konstruktion, Imagination und Inszenierung im touristischen Raum: Die Heurige in Wien aus sozial-und kulturanthropologischer Perspektive. Grin Verlag, München.

Robinson, J. (2006). The Oxford Companion to Wine. 3rd edition. Oxford University Press, USA.

Schiener, G. (2007). Die "klassischen" Heurigen und Buschenschenken im Burgenland und die Erwartungshaltung deren unterschiedlicher Besuchertypen, Master Thesis, Burgenland University of Applied Sciences.

Seidl, C. (2012) Land der Winzer, Der Standard, 25/26th October. Retrieved on 8/4/15 at http://derstandard.at/1350259333391/52-Prozent-Oesterreich-steht-besser-da-als-uebriges-Europa/

Sinhuber, B. F. (1996). Der Wiener Heurige. 1.200 Jahre Buschenschank.Coor. Amalthea,Verlag, Munchen.

Statistik Austria (2012). Tourismus Statistiken. Retrieved on March 26, 2014 at www.wien.gv.at/statistik/wirtschaft/tourismus/

Statistik Austria (2013). Statistics on Austrian Wine Industry. Retrieved on 8/6/15 at http://www.wien.gv.at/english/

Stöckl, A., Caseau, C. & Brouard, J. (2014). Traditional Wine Taverns and their hard Landing in the 21st Century-the Case of the Viennese Heurigen. Conference paper presented at Enometrics Conference, Lyon, France.

Strobl, M. (2002). Weinbau und Heurigenkultur in Wien. Analyse einer spezifischen Verbindung traditioneller Wiener Betriebstypen als Überlebensstrategie in der Großstadt, Diplomarbeit an der Wirtschaftsuniversität Wien.

Tagliabue, J. (1985). Scandal Over Poisoned Wine Embitters Village in Austria. *New York Times*, Aug. 2, 1985. Retrieved on 8/4/15 at http://www.nytimes.com/1985/08/02/world/scandal-over-poisoned-wine-embitters-village-in-austria.html

Teague, L. (2009). Is Grüner a Great Wine or a Groaner? *Food & Wine*. Retrieved on 8/7/15 at http://www.foodandwine.com/articles/is-grüner-a-great-wine-or-a-groaner

UNWTO. (2015). World Tourism Highlights – 2015 edition. Retrieved on 8/6/15 at http://www.e-unwto.org/doi/pdf/10.18111/9789284416899/

USDA (2014). Report on Austrian Wine. Retrieved on 8/4/15 at http://gain.fas.usda.gov/Recent%20GAIN%20Publications/Austrian%20Win e%202014_Vienna_Austria_2-3-2014.pdf

Vienna.Info. (2009). Vienna Brand Manual. Retrieved 18/5/15, http://b2b.wien.info/media/files-b2b/brand-manual.pdf

WienWein. (2014). Website of WienWein. Retrieved on 8/4/15 at http://www.wienwein.at/en/

Wien.Info (2014). The Viennese Heurigen. Retrieved on 8/4/15 at http://www.wien.info/en/shopping-wining-dining/wine/heurige-2

Chapter 3: CANADA – Creating Winter Wine Festivals in Niagara, Canada by Carman Cullen & Linda Bramble

Map & Photo Credits

Maps courtesy of the Wine Council of Ontario

Photos courtesy of the Grape Growers of Ontario and Carole & Roy Timm Photography

References

Canadian Wine and Grape Industry Fact Sheet. (March 2013) from http://www.winesofcanada.com/facts-canada.html/

CVA, (2014). *Canadian Wine and Grape Industry Fact Sheet*, Canadian Vintner's Association. from: http://www.canadianvintners.com/

CVA, (2013). *Canada's Wine Economy – Ripe Robust Remarkable*, Canadian Vintners Association, the Winery & Grower Alliance of Ontario, the BC Wine Institute and Winery Association of Nova Scotia. from: http://www.canadianvintners.com/

Goldenberg, S. (2013). Climate change will threaten wine production, study shows. *The Guardian*. from: http://www.theguardian.com/environment/2013/apr/08/climate-change-wine-production

Maclean's, (2014). *Wine in Canada*, Maclean's Magazine Special Edition, page 69.

National Geographic Travel (November 2014). from http://www.worldtravelguide.com/travel-gourmet/canada/ontario/niagara/niagara-ice-wine-festival.html

Pachereva, J.D, (November, 2014). Personal Interview

Rowland, W. (2014). *Niagara Icewine Festival*. from: http://www.worldtravelguide.com/travel-gourmet/canada/ontario/niagara/niagara-ice-wine-festival.html

Thompson, J. (November, 2014). Personal Interview

Twenty Valley's Winter WineFest. (November 2014). from http://www.20valley.ca/page/winter_winefest

VQA Ontario, (2014). from: www.vqaontario.ca/Appellations/NiagaraPeninsula

WGAO, (2013). from:
http://www.wgao.ca/uploads/Industry%20Newsletter%20-
%20April%202013.pdf

Chapter 4: CHINA - Building a Wine Destination from Scratch in Ningxia, China *by Wenxiao Zhang & Liz Thach*

Map and Photo Credits
China Map courtesy of L. Thach
Ningxia Map courtesy of Ningxia Bureau of Grape and Floriculture
Development
Photos courtesy of Ningxia Bureau of Grape and Floriculture Development and
L. Thach

References
Branigan, T. (2014). "China becomes biggest market for red wine, with 1.86bn
bottles sold in 2013", *The Guardian. Retrieved on 8/11/15 at*
http://www.theguardian.com/world/2014/jan/29/china-appetite-red-wine-
market-boom/
Cao, K. (2014). Presentation by Kailong Cao at International Wine Tourism
Seminar. Yinchuan, China. December 2014
Cao, K. (2015). Presentation by Kailong Cao to media. Yinchuan, China. March
2015
Changyu.com (2014). Website of Changyu. Retrieved on 8/12/15 at
http://www.changyu.com.cn:8189/
China Today. (2014). China City and Province: Ningxia Hui Autonomous
Region. *Chinatoday.com*. Retrieved on 8/18/15 at
http://www.chinatoday.com/city/ningxia.htm
China.org (2011). Helan Today. *China.org website*. Retrieved on 8/17/15 at
http://www.china.org.cn/travel/Ningxia/2011-01/11/content_21715104.htm.
ChinaCulture.org. (2003). Introduction to the oldest wine.
ChinaNewsDaily.com. Retrieved on 8/12/15 at
http://www.chinaculture.org/gb/en_curiosity/2005-01/26/content_65456.htm.
Clove Garden. (n.d.). Fat Choy - Black Moss. *Clove Garden Blog*. Retrieved on
8/17/15 at http://www.clovegarden.com/ingred/al_fatchz.html
Gastin, D. (2014). Understanding the China Syndrome. *Winestate Annual 2015*.
Dec. 17, 2014.
Getz. D. (2001). *Explore Wine Tourism: Management, Development &
Destinations*. NY: Cognizant Communications
Jiang Lu, (2015). Grape Wine Today in China. Presentation at UC Davis from
China Agricultural University in Beijing. 3/16/15. Retrieved on 8/12/15 at
http://confucius.ucdavis.edu/local_resources/js/JIangLU_Grape%20Wine%2
0Today%20in%20China%203-26%20UC%20Davis.pdf
Johnson, H & Robinson, J. (2013). "China." Chapter in 7th edition of *The World
Wine Atlas*. London: Octopus Publishing Co.
Lawrence, J. (2012). China to drink more white wine, says study. *Decanter*.
Retrieved on 8/12/15 at http://www.decanter.com/wine-news/chinese-to-

drink-more-white-wine-study-predicts-28569/

Lechmere, A. (2011). Chinese wine wins top honor at Decanter World Wine Awards. *Decanter*, Sept. 8, 2011. Retrieved on 8/17/15 at http://www.decanter.com/wine-news/chinese-wine-wins-top-honour-at-decanter-world-wine-awards-36689/.

Lyons, W. (2013). "Indulge in China's Latest Export". *Wall Street Journal*. Retrieved on 8/12/15 at http://www.wsj.com/articles/SB10001424127887323296504578396131833363780

Moselle, M. (2015). Ningxia Winemaker's Contest to shine spotlight on rising Chinese region. *Food & Drink*. Retrieved on 8/18/15 at http://www.scmp.com/lifestyle/food-drink/article/1831690/ningxia-winemakers-contest-shine-spotlight-rising-chinese.

OIV (2014). State of World Vitiviniculture situation. Paper presented at *37th World Congress of Vine and Wine*. Mendoza, Argentina. Nov. 10, 2014.

Robinson, J. (2012). Changyu, Cabernet Gernischt Blend 2011 Ningxia. *JancisRobinson.com*. Retrieved on 8/17/15 at http://www.jancisrobinson.com/articles/changyu-cabernet-gernischt-blend-2011-ningxia

Robinson, J. (2012a). Emma Gao - a story of wine today. *JancisRobinson.com*. http://www.jancisrobinson.com/articles/emma-gao-a-story-of-wine-today

Robinson, J. (2014). Australian triumphs in Ningxia. *JancisRobinson.com*. Retrieved on 8/17/15 at http://www.jancisrobinson.com/articles/australian-triumphs-in-ningxia

StatOIV (2014). *StatOIV Extracts*. Retrieved on 8/12/15 at http://www.oiv.int/oiv/info/enstatoivextracts2

Sylvia Wu, (2014). "Ningxia wine region: We've got your back, says the government", *Decanter- China*. Retrieved on 8/11/15 at https://www.decanterchina.com/en/?article=894

Thach, L. (2009). The Enchanting City of Turpan and the Uyghurs. *Wine Travel Stories Blog*. Retrieved on 8/12/15 at http://winetravelstories.blogspot.com/2009/09/enchanting-city-of-turpan-and-uyghurs.html

Thach, L. (2013). A Snapshot of wineries and Vineyards in the Ningxia Wine Region of China. *Wine Travel Stories Blog*. Retrieved on 8/12/15 at http://winetravelstories.blogspot.com/2013/12/a-snapshot-of-wineries-and-vineyards-in.html

Thach, L. (2014). "Chinese Wine Region Establishes Classification Modeled on Bordeaux's." *Wine Spectator,* Jan. 30, 2014. Retrieved on 8/11/15 at http://www.winespectator.com/webfeature/show/id/49539

The Economist. (2014). Playing the Long Game. *The Economist Intelligence Unit*. Retrieved on 8/17/15 at http://www.economistinsights.com/energy/analysis/playing-long-game

Wine China. (2014). Inside Industry. *Winechina.com – website of the China Alcoholic Drinks Association*. Retrieved on 8/18/15 at http://www.winechina.com/template/NewsEnSearch.aspx?caten=Inside%20Industry&caten2=China%20Wine%20Region

Zhuan Ti. (2014). "Ningxia Nurtures Big Plans For Wine", *Washington Post*. Retrieved on 8/11/15 at http://chinawatch.washingtonpost.com/2014/08/ningxia_nurtures_big_plans_for_wine/

Chapter 5: FRANCE (BORDEAUX) – "Toujours Bordeaux!" The Creation of a Cultural Wine Centerby Julien Cusin & Juliette Passebois Ducros

Photo & Map Credits
Maps courtesy of vin et vigne, www.vin-vigne.com
Photo of Margaux label courtesy of Château Margaux.
Photo of CWC architecture courtesy of Cité des Civilisations du Vin

References
Bonial. (2014). *Infographie Dynamique de la Consommation de Vin par Habitant Dans le Monde et les Régions Françaises*, October 2014.

Camp R.C. (1989). *Benchmarking: The Search for Industry Best Practices that Lead to Superior Performance*, Milwaukee, WI: ASQC Quality Press.

César, G. (2002). *Report on the Future of French Viticulture, July 2002*

CRT Aquitaine, (2013). Les Chiffres Clés de L'œnotourisme

Cusin J. et Passebois J. (2014), Les conditions d'une persistance appropriée dans un projet : le cas du centre culturel et touristique du vin, 23e conférences AIMS, Clermont Ferrand

CWC. (2015). Cite des Civilisations du Vin website. Available at: http://www.citedescivilisationsduvin.com/a-unique-concept-en.html

DGE (2014), Key Facts on Tourism, Direction Générale des Entreprises (http://www.entreprises.gouv.fr/etudes-et-statistiques/chiffres-cles-tourisme)

FEVS, (2013). Les exportations françaises de vin et spiritueux, Bilan Année 2013 et perspectives 2014.

FranceAgriMer, (2010). Note d'information de la filière vin France AgriMer, n° 168, June 2010

France AgriMer, (2013). La production de Vin en 2013

French Ministry of Economy, (2009), Création du conseil supérieur de l'oenotourisme, de la réflexion à la distinction de l'offre oenotourisme, Press kit.

Gonzalez S. (2011). "Bilbao and Barcelona 'in Motion'. How regeneration Models travel and Mutate in the global Flows of Policy Tourism", *Urban Studies*, 48(7):1397-1418.

Grodach C. (2008). "Looking Beyond Image and Tourism: The Role of Flagship Cultural Projects, Local Arts Development", *Planning, Practice & Research*, 23(4):495–516.

Haunschild P.R. & Miner A.S. (1997). "Modes of Inter-organizational Imitation: The Effects of Outcome Salience and Uncertainty", *Administrative Science Quarterly*, 42(3):472-500.

Inno'vin. (2006). Summary document - Bordeaux Aquitaine Inno'vin - 2006.

la revue du vin de France, 2012), Les régions viticoles : les vins de Bordeaux (http://www.larvf.com/,vins-bordeaux-aoc-medoc-saint-emilion-pomerol-graves-margaux-pessac-leognan-negociant,10355,4025372.asp)

Leve, J. (2015). Bordeaux Wine Production, Facts, Figures, Grapes, Vineyards. *Wine Cellar Insider.com*. Available at: http://www.thewinecellarinsider.com/wine-topics/bordeaux-wine-production-facts-figures-grapes-vineyards/

Lignon-Darmaillac S. (2009). *L'oenotourisme en France, Nouvelle valorisation des vignobles. Analyse et bilan*, Bordeaux, Feret édition.

Lynch, K. (1960). *The Image of the City*, The MIT Press.

Lynn G., Morone J. & Paulson A., (1996). "Marketing and Discontinuous Innovation: The Probe and Learn Process", *California Management Review*, 38(3):8-37.

Mommaas, H., (2004). "Cultural clusters and the post-industrial city: towards the remapping of urban cultural policy", *Urban Studies*, 41(3):507-532.

OIV, (2012). Note de conjoncture mondiale, Mars 2012

OpinionWay (2014), L'attractivité des métropoles Françaises, sondage OpinionWay pour foncière des régions, novembre 2014.

Plaza B., (2009). "Bilbao's Art Scene and the "Guggenheim effect" Revisited", *European Planning Studies*, 17(11):1711-1729.

Porter M.E., (1998). "Clusters and the New Economics of Competition", *Harvard Business Review*, November-December, 76(6):77-90.

Rimaud, Marie-Noël, (2011), Kaléidoscope des bonnes pratiques en matière d'oenotourisme, Les annales du vin et de ses marchés, Dereios Edition.

Schmitt B., (1999). "Experiential Marketing", *Journal of Marketing Management*, 15(1):53-67.

Staw B., (1976). "Knee-Deep in the Big Muddy: a Study of Escalating Commitment to a Course of Action", *Organizational Behavior and Human Performance*, 16(1):27-44.

Promos d Jour (2014). Bordeaux Wine Guide. Available at: http://www.promosdujour.com/guide-des-vins-de-bordeaux/ http://www.terroirs-france.com/

Vin et Société. (2011). La Filière Vin le Poids des Chiffres

Vin et société (2013), Chiffres clés de la filière vin (http://www.vinetsociete.fr/chiffres-cles)

Vin-Vigne, (2015). Available at: http://www.vin-vigne.com/vignoble/vin-bordeaux.html

Chapter 6: FRANCE (BURGUNDY) - Is Good Wine Enough? Place, Reputation, and Wine Tourism in Burgundy by Laurence Cogan, Steve Charters, Joanna Fountain, Claude Chapuis, & Benoît Lecat

Map and Photo Credits
Map of Burgundy – Courtesy of BIVB (http://www.bourgogne-wines.com/ Photos courtesy of S. Charters

References
Beaujolais Interprofessional Committee (2010). Tourism Survey Report. Unpublished study.

Beverland, M. (2006). "The 'real thing': Branding authenticity in the luxury wine trade." Journal of Business Research 59.2: 251-258.

BIVB. (2014). Website of Bourgogne Wine Board. Retrieved on 10/19/15 at
  http://www.bourgogne-wines.com/

BIVB. (2011). Passport to Burgundy Wines. Retrieved on 10/19/15 at
  http://www.bourgogne-wines.com/gallery_files/site/12881/13105/23596.pdf

Burgundy Tourism.com (2014). Practical Information. Website for Tourism in
  Burgundy. Retrieved on 10/19/15 at http://www.burgundy-tourism.com/

Charters, S. (2015) Experiencing wine tourism in Burgundy: An elite informant
  study. Contemporary Trends and Perspectives in Wine and Agrifood
  Management Conference; University of Salento, Italy, 16-17th January,
  Euromed Academy of Business

Charters, S., and Spielmann, N. (2014). The characteristics of strong territorial
  brands: The case of champagne. Journal of Business Research 67 pp. 1461-
  1467

Demossier, M. (2010) Wine Drinking Culture in France; A National Myth or a
  Modern Passion. Cardiff: University of Wales Press.

Dubrule, R. (2007). L'oenotourisme: Une valorisation des produits et du
  patrimoine vitivinicoles. Paris: Ministère de l'Agriculture et de la Pêche.

Frangin, B., (1994), Le guide du Beaujolais, Paris: Éditions La Manufacture.

Inter Beaujolais (2014). Beaujolais Facts and Figures. Discover Beaujolais
  Website. Retrieved on 10/19/15 at
  http://www.discoverbeaujolais.com/region/

Garrier, G., (2003), L'étonnante histoire du Beaujolais Nouveau, Paris:
  Larousse.

Hyde, K. (2013). A personal response to the provision of wine tourism in
  Burgundy. Unpublished Report: ESC Dijon

Lutun, A., (2001,) Beaujolais. Autour d'un vin, Paris: Flammarion.

OIV, (2015) Wine statistics. Accessed from
  http://www.oiv.int/oiv/info/enpublicationsstatistiques?lang=en (Accessed
  12th March 2015)

Pitiot, S., and Servant (2010) The Wines of Burgundy. Beaune: Collection
  Pierre Poupon

Young, A. (2013). The Top 10 Most Expensive Wines. TheDrinks
  Business.com. Retrieved on Sept. 13, 2015 at
  http://www.thedrinksbusiness.com/2013/09/the-top-10-most-expensive-
  wines/

Chapter 7: FRANCE (PROVENCE) – Pink Wine and Movie Stars: How the Provence Wine Trail Was Established by Coralie Haller, Sébastien Bede, Michel Couderc, & Francois Millo

Photo and Map Credits
Map and photos courtesy of the CIVP - Conseil Interprofessionnel des vins de
  Provence.

References
Atout France (2015). Wine Tourism – A strong Motivation to Visit a

Destination. Website of Atout France. Retrieved on Oct. 30, 2015 at http://atout-france.fr/content/oenotourisme

Camuto, R. (2014). Superstar Rosé: The inside story of Château Miraval, the Provence estate where Brad Pitt and Angelina Jolie pursue their wine-making dreams. *Wine Spectator*, June 30, 2014

Choisy, C. (1996), "The significance of viticultural tourism." *Espaces*, 140, 30-3.

CIVP - Conseil Interprofessionnel des vins de Provence. (2014), "Document économique: la filière du rosé".

Colombini, D. C. (2013), "Italian wine tourism and the web: A necessary wedding," *Wine Economics and Policy*, Vol. 2, n 2, p. 111-113.

Correia, P., Charters (2004), "Wine routes in Portugal: A case study of the Bairrada Wine Route," *Journal of wine research*, Vol. 15, n 1, p. 15-25.

Del Campo et al. (2010), "Wine tourism product clubs as a way to increase wine added value: the case of Spain," *International Journal of Wine Research*, p. 27-35.

Desplats, B.L. (1996). "A developed or developing tourism? Comparing the examples of armagnac and cognac," *Espaces*, 140, 34-42.

Faguet, J.-P. (2004), "Does decentralization increase government responsiveness to local needs?: Evidence from Bolivia," *Journal of public economics*, Vol. 88, n 3, p. 867-893.

FEVS (2013), Fédération des Exportateurs de Vins et Spiritueux de France.

Fischer, C. and Gil-Alana, L. A. (2009), "The nature of the relationship between international tourism and international trade: the case of German imports of Spanish wine," *Applied Economics*, Vol. 41, n° 11, p. 1345-1359.

France Agrimer (2013), "Key Figures of the French wine industry", http://www.franceagrimer.fr/filiere-vin-et-cidriculture/Vin/La-filiere-en-bref/La-production-de-vin-en-2013

Gatti, S., Incerti, F. (1997), "The wine routes as an instrument for the valorisation of typical products and rural areas", Proceedings of the 52nd EAAE Seminar, Prentice Hall International, (UK) Ltd, p.213-224.

Hall, C.M. et al., 1997. "Wine Tourism and Network Development in Australia and New Zealand: Review, Establishment and Prospects." *International Journal of Wine Marketing*, 9, pp.5–31.

Hojman, D. E. and Hunter-Jones, P. (2012), "Wine tourism: Chilean wine regions and routes", *Journal of Business Research*, Vol. 65, n° 1, p. 13-21.

http://www.fevs.com/fr/#/les-entreprises?idPage=2

Johson, H. & Robinson, J. (2013) Atlas Mondial du vin, 7ème Edition, Flammarion, Paris.

Mallon, P. (1996) "Wine and tourism: development in diversity." *Espaces* (Paris), 1996 No. 140, pp. 29-48.

Observatoire du rosé CIVP (2014), « Observatoire économique des marchés internationaux des vins rosés. » Study financed by CIVP and FranceAgriMer, realised by Agrex Consulting (2012-2014)

Organisation Internationale de la Vigne et du Vin, OIV, (2013), Point conjecture vitivinicole mondiale 2014.

PACA - Provence-Alpes-Côte d'Azur Tourism Board (2014). Site for Tourism Information. Retrieved on Oct. 30, 2015 at

http://www.infotourismepaca.fr/tendances-et-chiffres-cles/conjoncture/
Palanque, J.R. (1990). "Ligures, Celtes et Grecs". In Baratier, Edouard. *Histoire de la Provence*. Univers de la Toulouse: Editions Privat.
Wine Intelligence (2014), "Study on perception of consumers on rosé wine."

Chapter 8: ITALY - Città del Vino: A National Effort to Promote Wine Tourism in Italy by Roberta Capitello, Lara Agnoli, Ilenia Confente, Paolo Benvenuti & Iole Piscolla

Photo & Map Credits
Map of Italy PDO's: Authors' elaboration from Ismea (2013).
Map of Città del Vino: Authors' elaboration from the Città del Vino and Istat (2011)
Photos: Courtesy of Città del Vino (2014).

References

Cavicchi, A & Santini, C (eds) (2014). *Food and Wine Events in Europe: A Stakeholder Approach*. Routledge, Abingdon.
Città del Vino. (2015) Website of Citta Del Vino.,
http://www.cittadelvino.it/
Federdoc. (2014). I Vini Italiani a Denominazione D'Origine.,
http://www.federdoc.com/new/wp-content/uploads/2015/04/vqprd-ed2014.pdf
Ismea. (2013). Vini a denominazione di origine. Struttura, produzione e mercato.
http://www.ismea.it/flex/files/0/8/2/D.f8a5f07d28602ce2a8c8/2013_04_05_Report_vini_di_qualit__rev.1.pdf
Istat. (2011). Censimento dell'Industria e dei Servizi.
http://censimentoindustriaeservizi.istat.it
OIV. (2015). State of World Vitiviniculture situation.
http://www.oiv.int/oiv/info/enconjoncture?lang=en
ONT. (2012). Rapporto sul turismo 2011, National Tourism Observatory, Presidenza del Consiglio dei Ministri, Dipartimento per lo Sviluppo e la Competitività del Turismo.
http://www.ontit.it/opencms/export/sites/default/ont/it/documenti/files/ONT_2012-07-01_02836.pdf
ONT. (2014). ITALY MONITour, Statistics in tourism May 2014, *National Tourism Observatory*, Ministero dei beni e delle attività culturali e del turismo.
http://www.ontit.it/opencms/export/sites/default/ont/it/documenti/files/ONT_2014-05-16_03018.pdf
Unioncamere. (2009). Rapporto Nazionale sul Settore Vitivinicolo.
http://www.unioncamere.gov.it/P42A526C305S144/Rapporto-Nazionale-sul-settore-Vitivinicolo.htm
UNWTO. (2012). Global Report on Food Tourism, World Tourism Organization, Madrid.

http://dtxtq4w60xqpw.cloudfront.net/sites/all/files/pdf/global_report_on_foo
d_tourism.pdf

Chapter 9: NEW ZEALAND - Wine Tourism on an Isolated Island: The Intriguing Case of Waiheke, New Zealand by Lucy Baragwanath & Nick Lewis

Map and Photo Credits -
NZ Wine Map courtesy of Moran W. and McDowall C. Unpublished map of
New Zealand Wine Regions.
Waiheke Wine Map courtesy of Waiheke Winegrowers Association Inc.
Photos of Waiheke Island courtesy of waihekewine.co.nz and Cable Bay
Winery.

References
Baragwanath, L. (2010) 'The Waiheke Wine Project: overview of tourism, wine
and development on Waiheke Island', School of Environment Occasional
Publication 51, Auckland: University of Auckland.
Baragwanath, L. and Lewis, N. (2009) 'Waiheke Island Visitor Survey Report',
School of Environment Occasional Publication 50, Auckland: University of
Auckland.
Baragwanath, L. and Lewis, N. (2014) *Wine on Waiheke*, in Howland P. (editor)
Social, Cultural and Economic Impacts of Wine in Zealand. Routledge
Baragwanath, L., Lewis, N., Priestley, B.'Wine Tourism on Waiheke: Initial
lessons', paper presented at Romeo Bragato Conference, Napier 2009.
Deloitte and New Zealand Winegrowers (2014) *Vintage 2014 New Zealand
wine industry benchmarking survey*. Deloitte and New Zealand
Winegrowers, Auckland.
Hall, C. M. (2005). Rural Wine and Food Tourism Cluster and Network
Development. Rural Tourism and Sustainable Business. D. Hall, I.
Kirkpatrick and M. Mitchell. Clevedon, Channel View: 149-164.
NZWine. (2013). New Zealand Winegrowers Annual Report (2013). Available
at:
http://www.nzwine.com/assets/sm/upload/da/9i/te/eu/NZW_Annual_Report_
2013_web.pdf, Accessed 18 April 2015.
NZWine, (2014). New Zealand Winegrowers Annual Report (2014). Available
at:
http://www.nzwine.com/assets/sm/upload/b5/2j/rr/2n/NZW%20AR%202014
_web.pdf, Accessed 15 July 2015.
Kelly, G. (2009) 'Some Bordeaux blends from Waiheke Island, New Zealand'.
Available HTTP:
http://geoffkellywinereviews.co.nz/index.php?ArticleID=147 (accessed 22
March 2013).
Lewis, N. (2011). Mobilising Brands and Terroir in Champagne. In Charters S
(Ed.), *The Business of Champagne: A Delicate Balance*. Routledge.
Hayward, D. and Lewis, N. (2012) The Construction and Realisation of
Geographic Brand Rent in New Zealand Wine. *Urbani izziv*, 23(supplement

2), 2012, S49–S61

Lewis, N. (2014) Getting savvy and the flawed crisis of over-supply: A conceptual fix for New Zealand wine, in Howland P. (editor) *Social, Cultural and Economic Impacts of Wine in Zealand*. Routledge.

Lonely Planet (2012). 'Introducing Waiheke Island', Lonely Planet. Online. Available HTTP: http://www.lonelyplanet.com/new-zealand/auckland-region/waiheke-island> (accessed 29 April 2012).

McCann, S. (2014). *Statement of Evidence to the Environment Court.* Ministry of Justice http://www.justice.govt.nz/courts/environment-court/documents/susan-mccann (accessed 10 August 2015).

Ministry of Tourism (2009). *Wine Sector Profile.* Ministry of Tourism Urban Series B1.

Monin, P. (1992) *Waiheke Island: A History,* Palmerston North: The Dunmore Press

Moran W. and McDowall C. (2006) Unpublished map of New Zealand Wine Regions.

Picard, S. (2005) *Waiheke Island,* Auckland: RSVP.

Robinson, J. (2006) Ed, *Oxford Companion to Wine.* Oxford England: Oxford University Press.

Spratt, M. and Feldman, M. (2012) *Grape-a-hol: How big business is subverting artisan winemaking and the future of fine wine,* Indianapolis: Dog Ear Publishing.

Waiheke Winegrowers Waiheke Island of Wine website http://www.waihekewine.co.nz/ (accessed 18 April 2015)

Chapter 10: USA - Sonoma Sunshine: Learning to Collaborate for World Class Wine Tourism by Liz Thach & Michelle Mozell

Photo & Map Credits
US Map courtesy of K. Cavanaugh
Sonoma AVA Map courtesy of Sonoma County Winegrowers (sonomawinegrape.org)
Photos courtesy of Sonoma County Tourism (sonomacounty.com)

References
Adams, A. (2014). Record California Wine Grape Harvest. *Wines & Vines.* Retrieved on Nov. 28, 2014 at http://www.winesandvines.com/template.cfm?section=news&content=127944

Alley, L. (2007). "Researchers Uncover Identity of Historic California Grape: Spanish researchers solve mysteries surrounding the Mission variety and viticulture throughout the Americas". *Wine Spectator Online.* Retrieved Nov. 29, 2014 at http://www.winespectator.com/webfeature/show/id/Researchers-Uncover-Identity-of-Historic-California-Grape_3412

Bates, R.P., Mortensen, J. A., Lu, J. & Gray, D. J (1989). The History of Grapes in Florida and Grape Pioneers. Retrieved on Dec. 23, 2014 at http://mrec.ifas.ufl.edu/grapes/history/florida_grape_history.pdf.

Brager, D. (2014). 10th Annual Presentation of US Wine Consumer Trends. Santa Rosa, CA January 2014.

Carroll, S. (2014). New Economic Impact Data Indicates Sonoma County's Winegrowers and Winemakers Contribute $13.4 Billion to Local Economy. Retrieved on 1/6/15 at http://www.sonomawinegrape.org/press-releases/new-economic-impact-data-indicates-sonoma-countys-winegrowers-and-winemakers

Carroll, S. (2014). *Sonoma County To Become Nation's First 100% Sustainable Wine Region*. Retrieved 2 10, 2014, from Sonoma County Winegrowers: sonomawinegrape.org/files/SCW-Sustainability-Announcement.pdf

Cooke, G. M. & Vilas, E.P. (1989). California wineries: growth and change in a dynamic industry. *California Agriculture*. Retrieved on 1/6/15 at https://ucanr.edu/repositoryfiles/ca4302p4-62178.pdf

ctanetwork.com. (2014). About CTA. Retrieved 12/28/14 at http://www.ctanetwork.com/

County of Sonoma. (2014). Your Government. Retrieved on 2/20/15 at http://sonomacounty.ca.gov/

Discover California. (2014). Statistics. Retrieved Nov. 29, 2014 at http://www.discovercaliforniawines.com/media-trade/statistics/

Fort Ross-Seaview.Org. (2012). Our Story. Retrieved Nov. 29, 2014 at http://fortross-seaview.org/our-story/.

Hansen, H.J. & Miller, J. (1962). *Wild Oats in Eden: Sonoma County in the 19th Century*. Santa Rosa, CA; Santa Rosa Publisher.

Impact DataBank. (2014). Statistics on US Wine Industry in 2013. Impact Databank Newsletter, March 2014.

Johnson, R. & Bruwer, J. (2007). Regional brand image and perceived wine quality: the consumer perspective. *International Journal of Wine Business Research*, Vol. 19 Iss: 4, pp.276 – 297

LNV (Legendary Napa Valley). (2014). Destination Management Through Research. Retrieved on Nov. 29, 2014 at http://www.visitnapavalley.com/research_statistics.htm

McGinty, B. (1998). *Strong Wine: The Life and Legend of Agoston* Haraszthy. Santa Clara, CA: Stanford University Press.

NMWGA, 2012. History of Wine in New Mexico. Retrieved on Dec. 23, 2014 at http://nmwine.com/history/.

NVV (Napa Valley Vintners). (2014). History of Wine in the Napa Valley. Retrieved on Dec. 23, 2013 at: http://www.napavintners.com/napa_valley/history.asp.

OIV. (2015). Statistics. Retrieved on 2/20/15 at http://www.oiv.int/oiv/cms/index?lang=en

Pinney, T. (1989). *A History of Wine in America*. Berkeley, California: University of California Press.

SCT (Sonoma County Tourism). (2014) Statistics. Retrieved on Nov. 29, 2014 at http://www.sonomacounty.com/articles/media/statistics

SCV (Sonoma County Vintners). (2014). History of Wine in Sonoma County. Retrieved on Dec. 23, 2013 at: http://www.sonomawine.com/about-sonoma-county/history-of-sonoma-county-wine-country

SCV (Sonoma County Vintners). (2015). Sonoma County Vintners Press Kit. Retrieved on Nov. 29, 2015 at: http://www.sonomawine.com/press-room/sonoma-county-vintners-press-kit

SCW (Sonoma County Winegrowers). (2015). About. Retrieved on Nov. 19, 2015 at http://www.sonomawinegrape.org/about.

Sonoma-Style Service. (2013). Excerpt. Available at http://peoplefirstps.com/sonoma-style-service/

Thach, L. (2014). *Call of the Vine: Exploring Ten Famous Vineyards of Napa and Sonoma.* NY: Miranda Press.

TTB. (2014). Wine appellations of origin. Retrieved on Dec. 23, 2014 at http://www.ttb.gov/appellation/.

UNWTO. (2015). World Tourism Highlights – 2015 edition. Retrieved on 8/6/15 at http://www.e-unwto.org/doi/pdf/10.18111/9789284416899/

US Travel Association. (2014). *Travel is an Economic Engine.* Retrieved 4 26, 2014, from Travel Effect: traveleffect.com/sites/traveleffect.com/files/states/California_UST_TravelEffect_FactSheet_50states_2014-5.pdf

USDA. (2013). *Grape Crush Report Overview.* Retrieved 2 10, 2014, from USDA National Agricultural Statistics Service: http://www.nass.usda.gov/Statistics_by_State/California/Publications/Grape_Crush/Final/2013/201303gcbnarr.pdf

Wine Institute. (2014). Statistics. Available http://www.wineinstitute.org/

Wines & Vines. (2014). Number of Wineries Grows to 8,391 in North America. Retrieved on Dec. 23, 2014 at http://www.winesandvines.com/template.cfm?section=news&content=127266.

Wine Market Council. (2013). Research. Available at http://winemarketcouncil.com/

Chapter 11: ARGENTINA: From Piping Water to Piping Wine: The Zuccardi Wine Dynasty of Mendoza, Argentina by Jimena Estrella Orrego & Alejandro Gennari

Map and Photo Credits
Map of Argentina and Mendoza courtesy of Wines of Argentina
Photos of Zuccardi courtesy of Family Zuccardi Winery and L. Thach.

References
Casa del Visitante. (2014). Website of wine tourism experiences. Retrieved on 8/10/15 at http://www.casadelvisitante.com/

Caucasia Wine Thinking. (2015). Website. Retrieved on September 10, 2015 http://www.caucasia.com.ar/

Decanter. (2014). Discover the world's highest vineyards in the northern region of Argentina. Retrieved on Nov. 22, 2015 at : http://www.decanter.com/argentina-2014-coverage/discover-the-world-s-highest-vineyards-in-the-northern-region-of-argentina-30583/

INV (Instituto Nacional de Vitivinicultura). (2015). INV Website. Retrieved on

Nov. 22, 2015 at : http://www.inv.gov.ar/

Mateu, A.M. (2008). "La vitivinicultura mendocina entre 1870 y 1920: la génesis de un modelo centenario," book chapter in *El Vino y Sus Revoluciones*. Mendoza: EDIUNC

Thach, L. (2014). "Malbec and Empanadas: Wine Tourism in Argentina Healthy But Still Some Challenges. *Winebusiness.com*, March 25, 2014. Available at: http://www.winebusiness.com/news/?go=getArticle&dataid=130120

Thach, L. (2014). Delightful Day Long Visit to Zuccardi Family Winery - Mendoza, Argentina. *Wine Travel Stories*. Retrieved on 8/10/15 at http://winetravelstories.blogspot.com/2014/06/delightful-day-long-visit-to-zuccardi.html

UNWTO. (2015). World Tourism Highlights – 2015 edition. Retrieved on 8/6/15 at http://www.e-unwto.org/doi/pdf/10.18111/9789284416899/

Wine Enthusiast. (2014). Ten Best Wine Travel Destinations for 2014. Retrieved on 8/10/15 at http://www.winemag.com/February-2014/10-Best-Wine-Travel-Destinations-2014/index.php/slide/Mendoza--Argentina/cparticle/5

Wines of Argentina (2015). Key Data. Retrieved on 8/10/15 at http://www.winesofargentina.org/

Winter, P., Estrella-Orrego, J., Gennari, P., Eisenchlas, P., Ciardullo V. and Martin, D. "Collective or individual touristic strategies? An analysis of Argentinean oenoturism" en "Wine and Tourism: A value-added partnership for promoting regional economic cycles" Lena-Marie Lun, Axel Dreyer, Harald Pechlaner, Gunter Schamel (eds). EURAC book 62. EURAC Research. (2013)

Zuccardi.com (2015). Zuccardi Website. Retrieved on 8/10/15 at http://www.familiazuccardi.com/home_en.php.

Chapter 12: AUSTRALIA: Creating a Unique Wine Tourism Experience: The Case of Moorilla Estate in Tasmania, Australia by Marlene Pratt

Map and Photo Credits

Map of Australian wine regions courtesy of K. Cavanaugh
Map of Tasmania wine regions courtesy of Wine Tasmania.
Photos courtesy of Moorilla Estate

References

ABC News. (2015). MOFO 2015: Weather against Hobart's MONA FOMA festival but targets met. Retrieved 2/6, 2015, from http://www.abc.net.au/news/2015-01-19/weather-dampens-mona-foma-attendance-but-targets-met/6025924

Alant, K., & Bruwer, J. (2004). Wine tourism behavior in the context of a motivational framework for wine regions and cellar doors. Journal of Wine Research, 15(1), 27-37.

Australian Bureau of Statistics. (2012). Vineyard Estimates Australia, 2011-12

1329055002. Canberra: Australian Bureau of Statistics.

Australian Wine and Brandy Corporation. (2010). Geographical Indications. Retrieved 18 March, 2010, from http://www.wineaustralia.com/australia/Default.aspx?tabid=4467

Australian Wine online. (2006). Strategy 2025. Retrieved 23 September, 2007, from http://www.winetitles.com.au/awol/overview/strategy2025/

Cope, S. (2015). Ultimate Winery Experiences Australia. Retrieved 30/7, 2015, from http://www.ultimatewineryexperiences.com.au/about/

Cullen, S. (2013, 6 Mar). Govt backs wine expansion plan. ABC News. Retrieved from http://www.abc.net.au/news/2013-03-06/241m-boost-for-wine-industry/4555620

Dept of State Growth. (2014). The wine industry in Tasmania. A guide for investors. State of Tasmania: Department of State Growth,.

Getz, D. (2000). Explore Wine Tourism: Management, Development & Destinations. New York: Cognizant Communication Corporation.

Hall, C. M. (Ed.). (2003). Wine, Food, and Tourism Marketing. Binghamton: Hamworth Hospitality Press.

Hall, C. M., Johnson, G., Cambourne, B., Macionis, N., Mitchell, R., & Sharples, L. (2000). Wine tourism: an introduction. In C. M. Hall, L. Sharples, B. Cambourne & N. Macionis (Eds.), Wine tourism around the world: Development, management and markets (pp. 1-23). Oxford: Elsevier Science Ltd.

Hall, C. M., & Mitchell, R. (2008). Wine marketing a practical guide. Oxford: Elsevier Ltd.

Halliday, J. (2015). Wine Companion. Retrieved 12/2/15, 2015, from http://www.winecompanion.com.au/wineries/tasmania/southern-tasmania/moorilla-estate

Moorilla.com (2015). Website of Moorilla Estate. Retrieved on Nov. 28, 2015 at https://www.moorilla.com.au/

OIV. (2013). OIV Vine and Wine Outlook 2010-2011. Paris.

OIV. (2015). World vitiviniculture situation. Paris: OIV.

Sparks, B. (2007). Planning a wine tourism vacation? Factors that help to predict tourist behavioural intentions. Tourism Management, 28, 1180-1192.

Tourism Australia. (2013). Tourism Australia toasts new 'Best Of Wine' tourism initiative. Retrieved on Nov. 24, 2015 at http://www.tourism.australia.com/news/Media-Releases-2013-9190.aspx

Tourism Research Australia. (2010). Food and Wine tourism in Australia 2009. Canberra: Tourism Research Australia.

Tourism Tasmania. (2015a). MONA Visitor Profile. from Tourism Tasmania http://www.tourismtasmania.com.au/

Tourism Tasmania. (2015b). TASMANIAN VISITOR SURVEY AND TVS ANALYSER. Retrieved 12/2/2015 http://www.tourismtasmania.com.au/research/tvs

Walker, T. (2014). Vintage Tasmania: The complete book of Tasmanian wine. Dilston, Tasmania: Providore Island Tasmania.

White, C., & Thompson, M. (2009). Self determination theory and the wine club attribute formation process. Annals of Tourism Research, 36(4), 561-586.

Wine Australia. (2013a). Major Wine Regions of Australia.

http://www.wineaustralia.com/en/Search%20results.aspx?m=AnyTerm&q=
    wine%20regions&page=2

Wine Australia. (2013b). National Vintage Review – 2013. Retrieved 10/10,
    2014, from
    http://www.wineaustralia.com/en/Winefacts%20Landing/Grape%20and%20Wine%
    20Production/Winegrape%20crush%20and%20prices/Vintage%20In%20Review%2
    02013.aspx?ec_trk=followlist&ec_trk_data=Winegrape+crush+and+prices

Wine Australia. (2014). History, Australian viticulture. Retrieved 2/12, 2014,
    from http://www.wineaustralia.net.au/en-PH/history.aspx

Winebiz. (2014). Wine Industry Statistics. Retrieved 14/08/14, 2014, from
    http://www.winebiz.com.au/statistics/wineriestable25.asp

Winemakers Federation of Australia. (2002). National wine tourism strategic
    business plan 2002-2005: Winemakers Federation of Australia.

Winemakers Federation of Australia. (2011). Harnessing the tourism potential
    of wine and food in Australia: Winemakers Federation of Australia.

Winetitles. (2015). Wine Industry Statistics. Retrieved 30/11/15 from
    http://winetitles.com.au/statistics/

Chapter 13: FRANCE (BEAUJOLAIS) - Wine and Kids: Making Wine Tourism Work for Families in Beaujolais at Hameau Duboeuf Winery by Joanna Fountain & Laurence Cogan-Marie

Photo and Map Credits
Map of France Courtesy of K. Cavanaugh
Map of Beaujolais Region Courtesty of Inter Beaujolais
Photos Courtesy of J. Fountain

References
Allen, M. (2014, 12 March). The French Disneyland for Wines: Georges
    Duboeuf. *The Wine Profilers.*
    http://www.beaujolais.com/cartographie.php?page=DT1222852935&lang=e
    n&codej=anglais&time=20150201105607

Atout France (2010). *Tourisme et vin [Tourism and wine].*

Atout France (2014). *French wine tourist profile Survey.*

Backer, E. & Schänzel, H. (2012). The stress of the family holiday. In H.
    Schänzel, I. Yeoman & E. Backer (Eds.) *Family tourism: multidisciplinary
    perspectives* (105-124). Bristol: Channel View Publications.

Bruwer, J., & Alant, K. (2009). The hedonic nature of wine tourism
    consumption: an experiential view. *The Australian and New Zealand
    Grapegrower and Winemaker, 503,* 50-55.

Carr, N. (2011). *Children's and families' holiday experience.* London:
    Routledge.

Charters, S., Fountain, J., & Fish, N. (2009). 'You felt like lingering…'
    Experiencing 'real' service at the winery tasting room. *Journal of Travel
    Research, 48*(1), 122-134.

Cody, K., & Jackson, S. (2014). The contested terrain of alcohol sponsorship of
    sport in New Zealand. *International Review for the Sociology of Sport.*

DOI:1012690214526399.

Cogan-Marie, L. & Charters, S. (2014). Can wine tourism remedy poor wine marketing? The case of Beaujolais. *8th International Conference of the Academy of Wine Business Research (AWBR)*. Geisenheim University, Germany, 28-30 June.

Fountain, J., Schänzel, H., Stewart, E. & Körner, N. (forthcoming). Family experiences of visitor attractions in New Zealand: Differing opportunities for 'family time' and 'own time. *Annals of Leisure Research.*

Frochot, I. (2000). Wine tourism in France: A paradox? In C.M. Hall, L. Sharples, B. Cambourne & N. Macionis (Eds.), *Wine tourism around the world: Development, management and markets* (pp. 67–80), Oxford: Butterworth Heinemann.

Hameau Duboeuf (2013). *Twenty years Hameau Duboeuf.* http://www.hameauduvin.com/wp-content/uploads/Le-Hameau-Duboeuf-c%C3%A9l%C3%A8bre-ses-20-ans-ANGLAIS.pdf

Haurant, S. (2012, 16 June). A wine tasting trip in Bordeaux: for all the family. *The Guardian.* Accessed http://www.theguardian.com/travel/2012/jun/16/wine-tasting-bordeaux-family-holiday

Interprofession du Beaujolais (2009, 2010). *Wine tourist survey results.* (unpublished).

Inter Beaujolais (2014a). *Beaujolais Wine Tourism Guide Book.* http://www.beaujolais.com/page.php?page=DT1276870883&lang=fr&codej=france&time=20111018100001

Inter Beaujolais (2014b). Wine tourism. http://www.beaujolais.com/cartographie.php?page=DT1222852935&lang=en&codej=anglais&time=20150201105607

Kelly, B., Baur, L. A., Bauman, A. E., King, L., Chapman, K., & Smith, B. J. (2013). Views of children and parents on limiting unhealthy food, drink and alcohol sponsorship of elite and children's sports. *Public Health Nutrition, 16*(1), 130-135.

Lee, B., Graefe, A., & Burns, R. (2008). Family recreation: A study of visitors who travel with children. *World Leisure Journal, 50*(4), 259-267.

Lehto, X. Y., Choi, S., Lin, Y.-C., & MacDermid, S. M. (2009). Vacation and family functioning. *Annals of Tourism Research, 36*(3), 459-479.

McCabe, S., Joldersma, T., & Li, C. (2010). Understanding the benefits of social tourism: Linking participation to subjective well-being and quality of life. *International Journal of Tourism Research, 12*(6), 761-773.

Moscardo, G. (1999). *Making visitors mindful: principles for creating quality sustainable visitor experiences through effective communication.* Champaign: Sagamore Publishing.

Obrador, P. (2012). The place of the family in tourism research: Domesticity and thick sociality by the pool. *Annals of Tourism Research, 39*(1), 401-420.

OIV (Office International des Vins) (2014). *Wine statistics.* http://www.oiv.int/oiv/info/enpublicationsstatistiques

Orlin, M. (2014, 16 December). The magnificent man behind Beaujolais Nouveau. *Grape Collective.* https://grapecollective.com/articles/georges-duboeuf-beaujolais-nouveau-modest-celebrity-wine-world

Petrick, J. F., & Durko, A. M. (2013). Family and relationship benefits of travel experiences: A literature review. *Journal of Travel Research, 52*(6), 720-730.

Pikkemaat, B., Peters, M., Boksberger, P., & Secco, M. (2009). The staging of experiences in wine tourism. *Journal of Hospitality Marketing and Management, 18*, 237-253.

Pine, B. J., & Gilmore, J. H. (1999). *The experience economy: work is theatre & every business a stage.* Harvard: Harvard Business Press.

Prescott, B. (n.d.). *Beaujolais Nouveau – Hamlet of Wine and Georges Duboeuf.* http://www.intowine.com/duboeuf.html

Quadri-Felitti, D. L., & Fiore, A. M. (2013). Destination loyalty: Effects of wine tourists' experiences, memories, and satisfaction on intentions. *Tourism and Hospitality Research, 13*(1), 47-62.

Quadri-Felitti, D., & Fiore, A-M. (2012). Experience economy constructs as a framework for understanding wine tourism. *Journal of Vacation Marketing, 18*, 3-15.

Robinson, J. (2006) (ed). *The Oxford Companion to Wine, 3rd edition*, Oxford University Press

Schänzel, H. A., & Smith, K. A. (2014). The socialization of families away from home: Group dynamics and family functioning on holiday. *Leisure Sciences, 36*(2), 126-143. doi:10.1080/01490400.2013.857624

Schänzel, H. Z., Yeoman, I., & Backer, E. (eds.) (2012). *Family tourism: multidisciplinary perspectives.* Bristol: Channel View Publications.

Shaw, S. M. (1997). Controversies and contradictions in family leisure: An analysis of conflicting paradigms. *Journal of Leisure Research, 29*(1), 98-112.

Thach, L. & Olsen, J. (Nov. 2005). Putting the "family" back into family wineries: Or to be a kid-friendly tasting room – or not? *Vineyard & Winery Management.*

Tourism New Zealand (2014). *Tourism special interest: Wine tourism.* *http://www.tourismnewzealand.com/media/1132209/wine-tourism-profile.pdf*

UNWTO (2012).*Compendium of Tourism Statistics: Data 2006-2010.* UNWTO: Madrid.

Van Westering, J., & Niel, E. (2004). The organization of wine tourism in France: The involvement of the French public sector. In C.M. Hall (Ed.) *Wine, food and tourism marketing*, (pp. 35-47), Abington, OX: Haworth Hospitality Press.

Chapter 14: GREECE - Reclaiming a Lost Heritage at Gerovassiliou Winery (Ktima), Greece by Caroline Ritchie & Kostas Rotsios

Photo & Map Credits
Photos are courtesy of Ktima Gerovassiliou Winery

Map of Greek Wine Regions courtesy of All About Greek Wine and Lisa Stavropoulos.

Map of Northern Greek Wine Regions courtesy of The Association of the Wines of Northern Greece

References

Corigliano, M. A. and Mottironi, C. (2013) "Planning and Management of European Peripheral Territories through Multifuncitinality: the Case of Gastronomy Routes", in Costa, C., Panyik, E. & Buhalais (eds) *Trends in European Tourism Planning and Organization.* Channel View Publications, Bristol. pp33-47

Estreicher, S. K., (2006). *Wine from Neolithic Times to the 21st Century.* Algora Publishing, USA

FAS (2014). *Wine Annual Report and Statistics,* FAS Europe Offices

Hellenic Statistical Authority (ELSTAT), (2013) Statistical Information and Publications Division, Pireaus www.statistics.gr.

Hellenic Statistical Authority (ELSTAT), (2014). *Greece in Figures.* Statistical Information and Publications Division, Pireaus www.statistics.gr.

Isle, G. (2009). Seven Green Varietals to Know. *Food & Wine.* Retrieved on April 15, 2015 at http://www.foodandwine.com/articles/greek-wines-7-greek-varietals-to-know.

Johnson, H. and Robinson, J (2013). *The World Atlas of Wine, 7th edition.* London Mitchell Beazley

Johnson, J. (2006). *The Story of Wine,* 3rd Edition. London: Mitch

Jordan, B. (2000) "Greek revival; after a century of stagnation, Greece is rebuilding its reputation for fine wines". *Hotel & Restaurant Magazine* September pp 32-34

Karafolas, S. (2006) "Creating a nonprofit network of producers for the development of local culture and tourism: the case of the wine roads of Northern Greece". *Review of International Cooperation.* Vol 66 No 1. Pp36-43

Karafolas, S. (2007). "Wine Roads in Greece: A Cooperation for the Development of Local Tourism in Rural Areas". *Journal of Rural Cooperation.* Vol 35 No. 1 pp71-90

Ktima Gerovassiliou (2014) Accessed December 2014 http://www.gerovassiliou.gr/

Meloni, G and Swinnen, J. (2013) "The Political Economy of European Wine Regulations". *Journal of Wine Economics.* Vol 8, No. 3 pp 244-248

Mitchell, R. Carters, S. and Albrecht, J.A. (2012) "Cultural Systems and the Wine Tourism Product". *Annuls of Tourism* Vol. 39 No, 1 pp311-335

Papadaki, A. (2015). Personal communication with A Papadaki. April 17, 2015.

Rynning, C. (2012). Wines of Macedonia. Retrieved on 4/15/15 at http://www.grape-experiences.com/2012/06/the-wines-of-macedonia-will-they-conquer-your-palate-too-part-1/

Stevenson, T. (1997). The New Sotherby's Wine Encyclopedia. Dorling Kindersley, London

Thalassi (2010). All about Greek Wine. Thalassi Companies. Retrieved on April 15, 2015 at http://www.allaboutgreekwine.com/regions/macedonia.htm.

Timothy D. J. and Boyd, S. W. (2015). *Tourism and Trails; Cultural, Ecological and Management Issues* Channel View Publications Bristol.

Tsartas, P., Papatheodorou, A. and Vasileiou (2014). "Tourism Development and Policy in Greece" in Costa, C., Panyik, E. & Buhalis, D. (eds) *European Tourism Planning and Organization Systems; The EU member states.*

Channel View Publications, Bristol pp 295-306
Unwin, T., (1996). *Wine and the Vine.* Routledge, London

Chapter 15: PORTUGAL - Attracting Wine Tourists to Alentejo, Portugal: The Case of Herdade da Malhadinha Nova by Paulo Mendes

Map and Photo Credits:
Map of Portugal courtesy of Wines of Portugal
Photos courtesy of Herdade da Malhadinha Nova

References
Conde Nast (2014). Guide to Alentejo. Available at:
http://www.cntraveller.com/guides/europe/portugal/alentejo/where-to-stay
Económico (2015). Turismo fluvial no Douro bate recorde de passageiros em
2014 . Available at: http://economico.sapo.pt/noticias/turismo-fluvial-no-
douro-bate-recorde-de-passageiros-em-2014_211616.html
Instituto do Vinho e da Vinha. (2015). Website. Available at:
https://www.ivdp.pt/index.asp?idioma=2/
Johnson, J. & Robinson. (2013). *The World Atlas of Wine*, 7th Edition. UK:
Mitchell Beazley.
Jones, G. (2006). Past and Future Impacts of Climate Change on Wine
Quality. *3rd International Wine Business Research Conference*,
Montpellier, 6-7-8 July 2006 working paper. Available at:
http://academyofwinebusiness.com/wp-content/uploads/2010/05/Jones-
G.pdf
Marcus, K. (2014). Douro Masterpiece - Winemakers craft beautiful reds in
Portugal's classic 2011 vintage. *Wine Spectator*, June 30, 2014 Issue.
National Geographic Traveler (2014): Best Trips 2014.
http://travel.nationalgeographic.com/travel/best-trips-2014/#/chapel-of-
bones-alentejo-portugal_72690_600x450.jpg
NY Times (2010). The Slow Lane. Available at:
http://www.nytimes.com/2010/05/23/t-magazine/23well-portugal-
t.html?pagewanted=2&_r=0
Robinson, J. (1986). *Vines, Grapes & Wines,* UK: Mitchell Beazley
Robinson, J. (2006) Ed, *Oxford Companion to Wine*. Oxford England: Oxford
University Press.
Tourism of Portugal (2015). Website. Available at:
http://www.turismodeportugal.pt/
US Today (2014). Ten Best Wine Regions to Visit. Available at:
http://www.usatoday.com/story/travel/destinations/2014/08/12/readers-
choice-best-wine-region/13944365/
ViniPortugal (2015). Press Kit. Available at:
http://www.viniportugal.pt/en/Press/PressKit.
Wines of Portugal (2015). Information. Available at:
http://www.winesofportugal.info/

Chapter 16: SPAIN - Wine Tourism in a Time of Economic Crisis: The Success Story of Can Bonastre Winery in Spain by Agusti Casas & Ruben Garcia

Map and Photo Credits

Map of Spanish Wine Regions created by Tyk based on Image: Comunidades autónomas de España.svg created by Emilio Gómez Fernández. Modified by L. Thach

Photos Courtesy of Can Bonastre Wine Resort, A. Romeo & R. Huertas (chapter authors)

References

ACEVIN (2009). *Informe de visitants a bodegas asociadas a las Rutas del Vino en España, Año 2009*. Madrid (Spain): Ed. Secretaria de Estado de Turismo. (accessed on Juny, 2015).
http://www.wineroutesofspain.com/bd/archivos/archivo68.pdf

Cohen, E., & Ben-Nun, L. (2009). "The important dimensions of wine tourism experience from potential visitors perception", *Tourism & Hospitality Research*, 9(1), 20-31.

Dougherty, P.H. (2012). *The Geography of Wine: Regions, Terroir and Techniques*. NY: Springer Science & Business Media, 2012

Gázquez-Abad, J.C., Huertas-García, R., Vázquez-Gómez, D., & Casas Romeo, A. (2015). Drivers of Sustainability Strategies in Spain's Wine Tourism Industry. *Cornell Hospitality Quarterly*, 56(1) 106–117

Getz, D., & Brown, G. (2006). "Critical success factors for wine tourism regions: a demand analysis", *Tourism Management*, 27, 146-158.

INCAVI (2013). *Denominació d'Origen Penedes*. Barcelona (Spain): Ed. Generalitat de Catalunya. (accessed on Juny, 2015).
http://incavi.gencat.cat/ca/denominacions-origen-protegides/denominacions-origen/penedes

Marzo-Navarro, M., and M. Pedraja-Iglesias. 2009. Wine tourism development from the perspective of the potential tourist in Spain. *International Journal of Contemporary Hospitality Management* 21 (7): 816-35.

Marzo-Navarro, M., and M. Pedraja-Iglesias.2012. Critical factors of wine tourism: Incentives and barriers from the potential tourist's perspective. *International Journal of Contemporary Hospitality Management* 24 (2): 312-34.

Oral-B. (2015). Oral-B Commercials at Can Bonastre Winery. Available at:
https://www.youtube.com/watch?v=MXH69u-a4uA and
https://www.youtube.com/watch?v=25N-rycB_SQ.

Porter, M. E.(1980). *Competitive strategy: techniques for analyzing industries and competitors*. New York: Free Press.

Porter, M.E. (1985). *Competitive advantage: creating and sustaining superior performance*, New York: Free Press.

Ray, G., Barney, J. B., and Muhanna, W. A. (2004), "Capabilities, business processes, and competitive advantage: choosing the dependent variable in empirical tests of the resource-based view", *Strategic Management Journal*,

Vol. 25 No. 1, pp 23-37.

Scherrer, P., A. Alonso, and L. Sheridan. (2009). Expanding the destination image: Wine tourism in the Canary Islands. *International Journal of Tourism Research* 11:451-63.

Soler, R. & Valls, F. (2004). Les memòries del dissortat Josep Bonastre de Santa Magdalena, pagès de Masquefa (1787-1815). *Miscellanea Aqualatensia*, 11, 102-108.

Vargo, S. L., and Lusch, R. F. (2004). Evolving to a new dominant logic for marketing. *Journal of marketing*, 68(1), 1-17.

Index

Index

Index

Index

Index

Index

Index

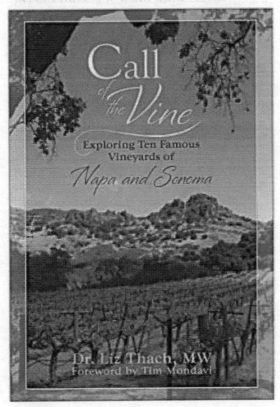